Zoophysiology Volume 38

Editors:
S.D. Bradshaw   W. Burggren
H.C. Heller   S. Ishii   H. Langer
G. Neuweiler   D.J. Randall

Springer
Berlin
Heidelberg
New York
Barcelona
Hong Kong
London
Milan
Paris
Singapore
Tokyo

Zoophysiology

Volumes already published in the series:

G. Kuchling

The Reproductive Biology of the Chelonia

With 87 Figures

Springer

Dr. Gerald Kuchling
The University of Western Australia
Dept. of Zoology
Perth, WA 6907
Australia

Cover illustration: By Guunolie Kuchling, modified from the picture book "Yakkinn the Swamp Tortoise"

ISBN 3-540-63013-9 Springer-Verlag Berlin Heidelberg New York
ISSN 0720-1842

Cover design: Design & Production GmbH, Heidelberg
Typesetting: Best-set Typesetter Ltd., Hong Kong

SPIN: 10087737   31/3137 – 5 4 3 2 1 0 – Printed on acid-free paper

Foreword

I had the good fortune to first meet Gerald Kuchling in 1985 when attending a conference held at the CNRS research centre, the Centre d'Etudes Biologiques des Animaux Sauvages (CEBAS), which is a wonderful scientific field station tucked away in the Forêt de Chizé, not far from the southwest town of Niort in France. The topic of the meeting, which included many invited overseas scientists, was "Physiological Regulations as Adaptive Mechanisms", and it was superbly organised by Ivan Assenmacher and Jean Boissin who together edited the important volume of papers published from the meeting. My stay in Chizé was, unfortunately, cut short as the wife of a very close colleague died on the night of the first day of the meeting and I had to return to Paris to assist with the funeral – but not before I had the chance to meet and talk with Gerald. He was presenting some of his work on a terrestrial tortoise (or turtle depending upon one's origins) – *Testudo hermanni* – a species that is now extremely rare in France but still relatively abundant in the then Yugoslavia. Gerald had been working in Yugoslavia, collecting blood samples from both males and females in an attempt to decipher details of its reproductive biology. Gerald and I conversed in French about the possibility of his coming to work in Australia, as his English was not fluent at that time, and I was interested in having someone work on the long-necked tortoise *Chelodina oblonga* which I was convinced belonged to a very ancient lineage of Gondwana relicts still surviving in the cool southwest region of Western Australia.

It took some time and organisation after that chance meeting before Gerald and his talented artiste wife, Guundie, first set foot in Australia – 1987 in fact and then only because Gerald was able to secure funding from the Austrian government to support an initial 12-month stay. Gerald completed his projected study of the long-neck and then became interested in Perth's other very interesting tortoise, the western swamp tortoise, *Pseudemydura umbrina*, which at that stage was the world's most endangered chelonian and Australia's most threatened vertebrate species. This amazing tortoise was first described as a new species by Ludwig Glauert in the 1950s after its discovery by a young schoolboy. It was Ernst Williams at the Harvard Museum who recognised that the "new" species was in fact another specimen of *Pseudemydura umbrina*, a tortoise that had been collected early in the nineteenth century in Perth but not seen again. A single specimen had been lodged in the Museum in Vienna and Ernst Williams recognised the new specimen as belonging to the same species. Gerald's interest in this same animal had long ago been kindled by seeing this rare specimen in the Vienna Museum as a child where he spent many long days as a budding herpetologist perusing and studying the Museum's vast collection!

Gerald applied for, and was granted, Australian citizenship and thus began his long study of and devotion to the western swamp tortoise which has led to its amazing recovery from the brink of almost certain extinction. There were fewer than 50 individuals known to be alive in 1987 and the small number of females being guarded by the Department of Conservation and Land Management (CALM) had not laid eggs for over 6 years. It was a chance discussion with another researcher working in my laboratory at that time, Bruno Colomb, that led to Gerald's use of ultrasound to visualise the state of the ovary and its developing follicles in this rare tortoise. He found that their gonads were developing each year, but that the animals lacked a critical pulse in food supply in spring needed to mature the follicles. Once this was understood it was only a matter of time before *Pseudemydura umbrina* began laying viable eggs and, to date, over 80 captive-bred young have been returned to the wild. Gerald's ultrasound technique has now been applied to many species of chelonians around the world and it is far preferable to established methods that expose female tortoises with growing eggs to the potentially harmful effects of X-rays. Gerald and Guundie's commitment to these tortoises goes even further and they have produced and published two highly successful books for children, based on the life of the western swamp tortoise, which have been published in English, German and French.

Let me conclude with a few words about Gerald's book. It has been gestating for a number of years, but it has been worth the wait! Gerald has written the book in English, a language that he now speaks and writes with consummate skill and precision, and it is a testament to him that he has been able to master this language so well. The book is a rare blend of detailed anatomical and physiological information on chelonians, coupled with highly practical considerations that lie at the heart of any successful conservation programme. It will be a highly valuable source book, not only for specialists interested in the reproductive physiology of chelonians, but also for herpetologists interested in conservation of rare and endangered species who realise that understanding the process of reproduction, and how it is normally achieved, is the key to any successful conservation programme.

Don Bradshaw

Perth, March 1998

Acknowledgements

I am indebted to Professor Don Bradshaw who encouraged me to write and finish this book, even though it has taken much longer than planned. The ideas expressed in this monograph were obtained in large part from my own research experience, but they were also profoundly influenced by my long association and collaboration with Prof. Bradshaw's eco-physiological research group at the University of Western Australia. As the editor responsible for this book, Prof. Bradshaw also provided critical comments on its contents and helped with the English language. Ms Liana Christensen made final corrections of the English.

Much of my personal research into the reproductive biology of chelonians was part of endangered species conservation programmes, in particular of the western swamp tortoise (*Pseudemydura umbrina*) programme in Western Australia. I thank the past and present members of the Western Swamp Tortoise Recovery Team for collaboration and help. I cannot name all of them, but I am especially indebted to Dr. Andrew Burbidge, who shared freely his knowledge of that species, and to Mr. Dean Burford for help with the monitoring of reproduction in the captive colony of *P. umbrina*. I thank my wife Guundie who was a tireless companion and helper during field work in various parts of the world and who provided technical assistance with ultra-sound scanning.

My research in various parts of the world has over the years been supported by grants from the Austrian government (Fonds zur Förderung der wissenschaftlichen Forschung), the Australian government (Endangered Species Programme and Australian Research Council), the Western Australian government, and by grants from WWF-Australia, Conservation International, British Chelonia Group and Bundesverband für fachgerechten Natur- und Artenschutz. I also thank my parents for their support during my years as a student and especially my father who instilled in me as a child my fascination for chelonians.

Contents

Chapter 1

Introduction

Chelonians (order Testudines or Chelonia) are shelled reptiles commonly referred to as turtles, tortoises and terrapins, but these terms may have different meanings in different countries. Many other languages apply a single term for the entire group, for example "tortue" in French, "Schildkröte" in German. It is a recent trend in the English literature to use the term turtle in a general way for all chelonians, whether marine, freshwater or terrestrial forms; many authors restrict the term tortoise to the terrestrial chelonians of the family Testudinidae; and terrapin is sometimes used for various hard-shelled freshwater turtles (Harless and Morlock 1979; Pritchard 1979a; Gaffney and Meylan 1988; Ernst and Barbour 1989; Cogger 1988). This usage is adopted throughout the book.

1.1 Phylogeny

Chelonians belong to the class Reptilia, the oldest vertebrate group which reproduces terrestrially and is fully independent from water. In contrast to the typical amphibian egg, which has a gelatinous cover and develops in water or in a damp place, typical reptilian eggs have a calcareous or parchment-like shell which retards moisture loss. The amnion, a fluid-filled compartment surrounding the embryo, sets reptilian eggs apart from even the most terrestrially adapted amphibian eggs; the amnion provides the aquatic environment which the embryo needs for development within the egg. This and a variety of other structural features, such as a tough, dry, scaly integument, established the supremacy of reptiles on land during the late Paleozoic and the Mesozoic period, when they were the dominant tetrapods.

Reptiles are classified on the basis of their skull structure, particularly the presence and number of openings in the temporal region of the bony skull. The chelonian skull has no openings in the temporal region and the order Testudines belongs to the subclass Anapsida in which the earliest stem reptiles or Cotylosauria were often included. Gaffney and Meylan (1988) redefined the subclass Anapsida to comprise the Testudines and the Permian Captorhinidae; all other living reptiles and the birds belong to the Diapsida, a sister group of the Anapsida with which they form the Sauropsida. Mammals evolved from the Synapsida, a sister group of the Sauropsida. Synapsida and Sauropsida together form the Amniota. The concept that mammals branched off before the divergence between chelonians and birds occurred is supported by protein sequence data

(Caspers et al. 1996). Recently, chelonian relationship became a hot topic in phylogeny, with Lee (1996, 1997a) considering the pareiasaurs ancestral to chelonians and placing them into the Parareptilia, a sister group to the Eureptilia. Contrary to this view, Debraga and Rieppel (1997) suggested on morphological grounds that chelonians are advanced diapsid reptiles (Eureptilia), a concept refuted by Wilkinson et al. (1997) and Lee (1997b), but supported again by molecular data (Platz and Conlon 1997). Thus, there is still no generally accepted concept of the origin and higher phylogenetic relationships of the chelonians and their position relative to mammals, birds and the remaining reptiles. The majority of phylogenists currently favour the concept presented in Fig. 1.1, which makes chelonians an important group to determine the basal condition and early transformations of many amniote characters.

In contrast to the classes Mammalia and Aves, the class Reptilia is not a monophyletic, homogenous systematic entity. Cladistic schemes would not permit the term "reptile" which represents a grade rather than a clade, but the retention of the concept of reptiles is still useful for the understanding of analogies and the analysis of past functional patterns (Crews and Gans 1992). The common ancestors of all amniotes (reptiles, birds and mammals) were the stem reptiles which gave rise to the Sauropsida and Synapsida. The order Testudines is an ancient group which goes back to the early period of amniote evolution. Based on phylogenetic antiquity, as well as skeletal characteristics, the chelonians are more closely related to the stem reptiles than any other living order of the Amniota. The reptile-like fossils that later gave rise to the mammals presumably left the basic "reptilian" line near the origin of the chelonians. This phylogenetic position (which is not undisputed; see above) makes exploration of chelonian biological traits of particular interest. One might assume that some of the primitive conditions retained by chelonians represent the stem reptile state. Aside from a drastic modification of the skeleton, chelonians have remained quite generalised in most other aspects of their anatomy, physiology and behaviour.

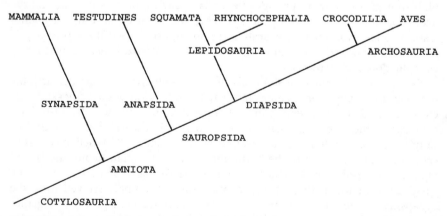

Fig. 1.1. Phylogeny of amniota with hypothetical placement of the chelonia according to Gaffney and Meylan (1988), Lee (1996, 1997a,b) and Wilkinson et al. (1997)

1.1.1 The Chelonian Bauplan

The anatomy of the chelonian shell is unique among tetrapods, and makes turtles one of the most clearly defined of all vertebrate orders. The bony shell is a composite of dermal bones and the endochondral axial skeleton, and consists of a dorsal carapace and a ventral plastron; the carapace is formed from costal bones with fused ribs, neural bones with fused thoracic vertebrae, and marginal bones. At the lateral margins the carapace is articulated to the plastron which is formed from interclavicle, clavicle, and three to five paired bones sutured together. The dermal, bony armour – and generally also the epidermal, horny surface cover – are subdivided into mosaics of discrete geometric shapes that form characteristic, though not congruous, patterns of high phyletic stability.

An association of dermal bones and endoskeletal elements is typical in the cranial skeleton of vertebrates, but very rare postcranially. Most remarkable in chelonians is the position of the ribs relative to the limbs and girdles: the carapace with the fused ribs overlaps the pectoral and the pelvic girdles anteriorly, posteriorly and laterally. Therefore, in strong contrast to other tetrapods, the shoulder girdle is enclosed by the ribs and situated ventrally, deep to the axial elements (Fig. 1.2). The chelonian body form offers armoured resistance to attack by predators, but the trade-off is reduced speed and agility.

The evolutionary history of the turtle body plan is obscure. Hypothetical ancestors are often constructed in the light of assumptions about the selective advantage of the shell as a protective adaptation. *Eunotosaurus africanus* from the middle Permian of South Africa was sometimes cited as the "missing link" between the Chelonia and more primitive cotylosaurs. The ribs of this reptile are expanded along their shafts into broad plates. However, since no dermal bones are associated with the axial skeleton and since the shoulder girdle takes the

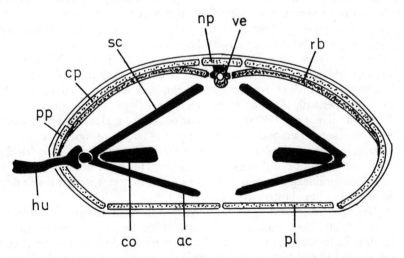

Fig. 1.2. Cross section through a generalised chelonian shell in the area of the shoulder girdle, showing relative position of skeletal elements. *ac* Acromion; *co* coracoid; *cp* costal plate; *hu* humerus; *np* neural plate; *pl* plastral plate; *pp* peripheral plate; *rb* rib; *sc* scapula; *ve* vertebra

standard tetrapod position, external to the ribs, it is now thought that *Eunotosaurus* is not in the line of chelonian evolution (Gaffney 1984).

Lee (1996, 1997a) suggested that pareiasaurs, a group of large herbivorous anapsid reptiles, are ancestral to chelonians and their nearest relatives. Evolutionary trends within pareiasaurs, such as the elaboration of the dermal armour, shortening and stiffening of the presacral region, and increasing reliance on limb-driven as opposed to axial-driven locomotion, shed light on how the distinct carapace of chelonians might have evolved through "correlated progression". The dermal armour may have first arisen to perform a supporting function and, as an "exaptation" *sensu* Gould and Vrba (1982), may have only later become co-opted for protection.

Embryological studies of the generation of the chelonian morphology gave some insight into the patterns responsible for the ontogeny of the turtle body plan. Carapace development seems to be dependent on epithelial–mesenchymal interaction in the body wall, the carapacial ridge, which may also have a causal connection to the placement of the ribs. This novel interaction in the body wall may have been instrumental in the evolutionary transition from the typical tetrapod arrangement of the trunk to that of the chelonian (A.C. Burke 1991). Unfortunately, it is not yet possible to correlate developmental and fossil evidence to establish how the processes described by Burke were reflected in phylogenetic change.

1.1.2 Fossil History and Evolution

The monophyly of the Testudines has never been questioned and is firmly established. The known history of the order Testudines begins in the Triassic with the fossil *Proganochelys*, which already had a full carapace and plastron. Its cranial anatomy links it to the Permian Captorhinidae, which have been identified as a sister group of the Testudines (Gaffney and Meeker 1983). Lee (1996), however, considers the evidence linking the chelonians to the Captorhinidae to be weak. Captorhinidae do not have the highly derived trunk morphology by which we recognise chelonians. No intermediate morphologies are represented in the fossil record between *Proganochelys* and the Captorhinidae. Lee (1997a) relates chelonians to a clade of pareiasaurs which exhibit otherwise uniquely chelonian features such as a rigid covering of dermal armour over the entire dorsal region, expanded flattened ribs, cylindrical scapula blade and some peculiarities of the limbs. However, because of the extreme morphological leap between chelonians and other tetrapods, it is still difficult to imagine functional intermediates.

The Proganochelyidea are the most primitive of all turtles and are a sister group of all other known chelonians, which together form the Casichelydia (Gaffney and Meylan 1988). The Casichelydia are divided into two extant suborders, the Pleurodira which flex the neck laterally to retract the head and the Cryptodira which use a sagittal flexure of the neck. Character analysis (Gaffney et al. 1991) supports the monophyly of both suborders.

Chelonians have a relatively rich fossil record and obviously reached a greater diversity in former geologic eras than they do today. It has been suggested that the oldest chelonians, the Proganochelyidae, were probably marsh dwelling and that the fully aquatic preference of many turtles is a secondary modification. However, the move to the freshwater medium was a rather early development and the majority of both modern and extinct turtle genera have or had a freshwater habitat. Truly terrestrial chelonians appeared in the Upper Cretaceous in different ancient families – the assumed terrestrial life style is postulated on several morphological parallels to modern terrestrial forms. The terrestrial tortoises of the family Testudinidae do not appear in the fossil record until the mid-Eocene and reached their greatest abundance and diversity in the Pliocene. Tortoises (Testudinidae) are relatively closely related to the Bataguridae and Emydidae which have also given rise to several other terrestrial lines, e.g. the box turtles of the genus *Terrapene* (Emydidae) and the genera *Pyxidea, Cuora* and *Rhinoclemmys* (Bataguridae). Turtles entered the marine environment at an early stage, in the Jurassic and Cretaceous, and it appears that marine lineages evolved independently at least three times. In addition, some of the early side-necks of the Podocnemidae (of which the living species are freshwater forms) were marine (Pritchard 1979b).

Figure 1.3 shows the relationships of the living turtle families according to Shaffer et al. (1997). Most groups retained the typical chelonian shell of dermal bony plates which are sutured together and covered by a mosaic of epidermal horny scutes; this extremely conservative character has been little changed for about 200 million years. A reduction in the size and number of bones in the shell occurred in the families Dermochelyidae and Trionychidae, which have also lost

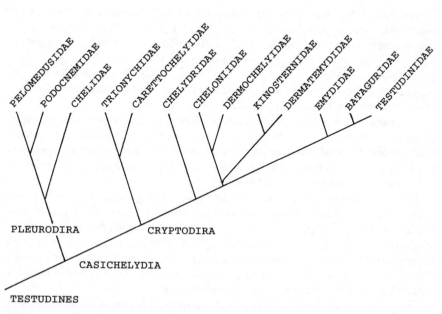

Fig. 1.3. Relationship of living chelonian families according to Shaffer et al. (1997)

the horny covering of scutes; instead, they have a tough, leathery skin. These turtles are referred to as leatherbacks and softshells. The well-developed bony shell of the monotypic family Carettochelyidae is covered with a leathery skin which resembles that of the Trionychidae.

All chelonians can walk or crawl on the land, all can swim, and most can move slowly along the bottom of bodies of water (Walker 1979). The shape of the shell and the limbs is adapted to the medium through or on which they travel. Shell shapes range from oval, highly domed carapaces of some terrestrial chelonians through heart-and-lyre shaped, streamlined shells of sea turtles to rounded, flat-tened carapaces of soft-shelled turtles. The terrestrial tortoises (Testudinidae) have evolved elephantine hind limbs and club-shaped front legs, which help to support them. The feet of semiaquatic turtles have independently movable toes with claws and with various degrees of webbing between them. In the vast major-ity of aquatic chelonians, the limbs represent a compromise between the needs of swimming and of walking. Generally, the more webbing, the more aquatic the turtle. In many swimming forms the hind limbs are longer, more powerful, and have more extensive webbing than the forelimbs, and they generate the principal propulsive force. However, in some of the most aquatic chelonians, the forelimbs provide the principal propulsive force. The forelimbs of the marine species (Cheloniidae and Dermochelyidae) and of *Carettochelys* are modified as paddle-like flippers with which they "fly" through the water.

1.1.3 Systematics and Biogeography

Chelonians are found on all continents except Antarctica, and in all oceans except where the water is permanently too cold. This is only a short overview of the biogeography of recent families. Detailed discussions can be found in Ernst and Barbour (1989), Iverson (1992a) and Pritchard (1979b). New turtle species and genera keep being discovered and opinions on chelonian systematics diverge. The following numbers of genera and species are, therefore, disputable and only given as an indication of the diversity of the different groups.

Some sea turtles have vast distributions. The leatherback turtle *Dermochelys coriacea*, representing monotypically the family Dermochelyidae, occurs worldwide in tropical, subtropical and temperate seas. Four of the six species of the hard-shelled sea turtles of the family Cheloniidae have a circumtropical to subtropical distribution around the globe.

The two closely related families Emydidae (10 genera, 35 species) and Bataguridae (23 genera, 59 species) were previously assigned the status of sub-families in a single family Emydidae (*sensu lato*) and include the majority of the freshwater turtles of the northern hemisphere. Emydidae reach their greatest diversity in eastern North America and Bataguridae in Southeast Asia. They are absent in subSaharan Africa and in Australia. In South America, both Emydidae and Bataguridae are represented by a single genus.

The terrestrial Testudinidae (12 genera, 50 species) are found on all continents except Australia (and Antarctica). Modern tortoises reach their greatest diversity in Africa, south of the Sahara. Testudinidae float well in water and are able to cross oceanic barriers: they reached both offshore and remote oceanic islands.

The soft-shelled turtle family Trionychidae harbours two subfamilies, the Cyclanorbinae (3 genera, 6 species) and the Trionychinae (11 genera, 17 species). Today, Trionychidae live in Africa, Asia, the Indo-Australian archipelago and North America. Fossils are also known from Europe, Australia and South America. The related, monotypic family Carettochelyidae is today restricted to New Guinea and northern Australia.

Kinosternidae is a New World family of small- to medium-sized semiaquatic turtles ranging from Canada to South America. Two subfamilies are recognised, the Staurotypinae (2 genera, 3 species) and the Kinosterninae (1 genus, 19 species). The related family Dermatemydidae has only one living representative, the Central American river turtle *Dermatemys mawii*.

The family Chelydridae includes two genera in America: the large and savage snapping turtles of the genus *Chelydra* (3 species; Phillips et al. 1996), which ranges from southern Canada to Ecuador; and *Macroclemys temminckii* restricted to the USA. According to Gaffney and Meylan (1988) the Asiatic big-headed turtle *Platysternon megacephalum* also belongs to this family; it was long thought to represent a monotypic family on its own.

All families above belong to the suborder Cryptodira. The suborder Pleurodira contains the remaining three families, the Podocnemidae, Pelomedusidae and the Chelidae. The Podocnemidae and Pelomedusidae were long considered to represent only subfamilies (Podocneminae and Pelomedusinae) of a larger family Pelomedusidae, but recent molecular, morphological and paleontological studies attributed them separate family status (Shaffer et al. 1997). Today, Pleurodira are largely confined to the southern continents.

Pelomedusidae (2 genera, 17 species) occur in subSaharan Africa including Madagascar and other Indian Ocean islands. The Podocnemidae have two extant genera with seven species in tropical South America and one monotypic genus in Madagascar, but fossils of this subfamily are also known from Europe, Asia, mainland Africa and North America.

Chelidae are known from South America and the Australian region and are the only chelonian group of obviously Gondwanean origin. Gaffney and Meylan (1988) recognised two subfamilies, the monotypic Pseudemidurinae of which only a few individuals survive in the south-western corner of Australia and its sister group, the Chelinae (10 genera, 40 species), which include all other Australasian and the South American chelid taxa. Recent evidence, however, suggests that both the Australasian and South American lineages may be monophyletic sister groups (Shaffer et al. 1997).

1.1.4 Body Size

The encasement of the chelonian body inside a box-like armour sets relatively rigid limits to the space which is available for food intake, breathing volume, energy and water storage, as well as reproductive output (size and number of eggs). To increase space, an obvious solution is to grow larger. Body size plays an important role in reproductive traits: interspecifically, larger species can lay more or larger eggs per clutch than smaller ones; intraspecifically, large females of several chelonian species have a higher reproductive success and higher quality offspring than smaller ones (see Chaps. 8 and 9).

Larger size improves the volume to surface ratio of the animal, but has the drawback that agility is lost, in particular in terrestrial tortoises that have to carry their armour and body mass. Aquatic turtles which swim and float most of their life do not lose as much agility by increasing body size and mass. Sea turtles, as an ecological group, are all large-bodied. All the larger species of freshwater turtles live in large water bodies like rivers and lakes and are highly aquatic. The largest living chelonian is the marine leatherback turtle *Dermochelys coriacea*, reaching a shell length of 244 cm and possibly a body mass of 867 kg (Ernst and Barbour 1989). The fossil marine *Archelon ischyros* of the Upper Cretaceous reached a total length (including head) of 3.5 m and spanned 5 m between its flippers (Obst 1985).

Some terrestrial tortoise species grow much bigger than most of their relatives. The largest living tortoises are the giant tortoises of the Galapagos Islands in the Pacific (*Geochelone nigra*, up to 134 cm carapace length and 263 kg: Swingland 1989a; but up to 400 kg in captivity: Pritchard 1996) and of the Aldabra Atoll in the Indian Ocean (*Geochelone gigantea*, up to 105 cm carapace length and 250 kg: Swingland 1989b). Some of the continental species also reach respectable sizes: *Geochelone sulcata* of the African Sahel zone, 76 cm carapace length, and the South Asian *Manouria emys*, 60 cm (Ernst and Barbour 1989).

It is intriguing that the only tortoises found on oceanic islands are gigantic ones. Before sailors and settlers in the eighteenth and nineteenth centuries eliminated them, giant tortoises were found on dozens of islands in the western Indian Ocean and they are still found on most of the larger Galapagos Islands in the east Pacific Ocean. Gigantism seems to be a pre-adaptation for island life rather than a result of it, although we do not know the size of the first colonising tortoises. However, a large tortoise with its considerable storage capacity of energy and water is much more likely to survive passive flotation across many hundred kilometres of ocean than a small one. Another indication that gigantism of tortoises may be a pre-adaptation for, rather than a consequence of, life on oceanic islands are fossils of several large tortoises from continents and from Madagascar (which, biogeographically, is a small continent rather than an island). The largest known terrestrial tortoise ever was the Pleistocene *Geochelone sivalensis* (also known as *Colossochelys atlas*) from southern Asia, which had a carapace length of up to 2 m (Pritchard 1979a, 1996).

At the other end of the scale, the smallest chelonians are around the 100 g mark: in the speckled cape tortoise *Homopus signatus* males reach 85 mm shell length and 90 g and females 95 mm and 140 g (Boycott and Bourquin 1988); *Clemmys muhlenbergii* and *Kinosternon depressum* both reach 115 mm shell length and may be the smallest freshwater turtles (Ernst and Barbour 1989). Thus, living chelonians cover a mass range with a factor of more than one thousand.

1.2 Chelonian Reproductive Traits Compared with Other Reptiles

Reproduction in the class Reptilia exhibits a wide variety of traits: squamates (lizards and snakes) are mostly oviparous, but have evolved ovoviviparity and viviparity many times (Shine 1985); parthenogenesis also occurs in squamates (Crews 1989) and highly developed parental care occurs in crocodiles and some squamates (Shine 1988). Chelonians are oviparous and show little trend towards the various other traits found in reptiles. For example, parental care in the sense of parental behaviour enhancing the survival chances of offspring, once the eggs have been laid and nesting has finished, is uncommon in chelonians, although not totally lacking. Surprising is the total lack of traits in chelonians towards viviparity.

The universality of internal fertilisation in reptiles readily permits the evolution of viviparity. However, chelonians (as well as crocodiles and birds) are oviparous and show no trend to retain eggs and carry embryos for a longer time than needed to build the egg shell. Webb and Cooper-Preston (1989) suggest that turtle and crocodile eggs are specialised for early independence in a way that would make viviparity difficult to evolve. According to this theory the less specialised respiratory adaptations of lizard and snake eggs are less constraining. Although there seems to be no other obvious impediment to egg retention in turtles that would not also apply to squamates, different respiratory specialisation alone is a weak explanation for the lack of trends to viviparity.

Oviparity seems especially maladaptive for the highly aquatic sea turtles. Eggs, hatchlings and nesting females are readily exploited by terrestrial predators. These hazards would be completely avoided or much reduced if the young were born at sea, as with most sea snakes. The first steps toward viviparity could be taken by any turtle population for which a briefer period of development on the beach resulted in reduced loss of the young. The result would be decreasing periods of vulnerability for the eggs, for as long as benefits exceed costs. If such conditions persisted for an appreciable period of evolutionary time, viviparity would result (Williams 1992).

Viviparity in sea turtles may not have evolved because the main disadvantages of their oviparity do not depend on the duration of egg incubation; females are exposed to terrestrial hazards during laying, regardless of the time the embryos need to develop; losses of eggs may be heaviest immediately after nest construction when visual and olfactory cues are most obvious to predators, and the emerging young must scramble across the beach to the sea, at which time they are taken in large numbers by predatory birds, mammals and crabs, regardless of the incubation time. The intermittent stages demanded by the theory of natural selection (Williams 1992) may never completely compensate the disadvantage to the female in carrying the embryos for a longer time (e.g. reduced clutch frequency and, therefore, reduced number of eggs per nesting season). It may also hold for other chelonians that most threats to the progeny occur immediately after eggs are laid and when hatchlings disperse, with egg incubation itself being a minor hazard and a reduction of its duration without strong selective advantage.

1.3 Anatomy of the Sexual Organs

The sexual organs of vertebrates have a close anatomical relationship to the excretory system, together with which they form the urogenital system. The ontogeny, phylogeny and anatomy of the urogenital system of reptiles has been reviewed by Fox (1977). The gross anatomy of the urogenital system of a turtle – the European pond turtle *Emys orbicularis* – was described in detail relatively early by Bojanus (1819).

The paired male gonads, the testes, produce spermatozoa which leave the body via special ducts. They "swim" in a fluid produced by the epithelium of the spermiducts, making up the semen. All reptiles have internal fertilisation, in contrast to most amphibians where eggs are fertilised during spawning in water outside the body. During copulation, male chelonians insert a penis into the cloaca of the female to transfer semen into the female genital tract. The female gonads, the paired ovaries, produce ova which are set free into the body cavity during ovulation and borne out of the body through the oviducts. Fertilisation, the fusion of egg and sperm, has to take place in the proximal part of the oviduct before the secondary egg membranes are formed by oviducal glands.

1.3.1 Males

The testes of reptiles are similar overall to those of mammals and birds, and they are vascularised by spermatic ateries and veins (Fox 1977). The yellow, oval or elongated testes of adult chelonians are loosely bound to the kidneys by the mesorchium, the latter extending mediolaterally over the epididymis and kidney and being continuous with the body wall and the dorsal mesentery. The tunica albuginea at the outside of the testis contains blood vessels and encases the convoluted seminiferous tubules which are interspersed with interstitial cells and blood vessels. The interstitial Leydig cells are the main site of sexual steroid hormone synthesis in the testis. The seminiferous tubules end in the ductuli efferentes which form, outside the testis, a rete testis which is connected to the ductuli epididymides. The ductuli epididymides join into the long, convoluted ductus epididymidis which leads to the rather short vas deferens (Fig. 1.4). The microscopic anatomy of the seminiferous tubules with the germinal epithelium, of the interstitium of the testis and of the epididymis will be treated in Chapter 3.

Septae divide the cloaca of chelonians, at least partly, into an upper and a lower chamber. The vasa deferentia and ureters of each side open into the lower chamber which is referred to as the sinus urogenitalis. The unpaired penis is formed out of two longitudinal ridges forming a groove in the ventral floor of the cloaca. It is completely contained within the cloaca when in the relaxed state (Fig. 1.5). Each ridge is composed of a coelomic canal and an erectile tissue, the corpus cavernosum (or corpus spongiosum). Erection is caused by tumescence, in contrast to penile erection in snakes and lizards which is essentially a process of eversion. During erection the corpora cavernosa fill with blood and become

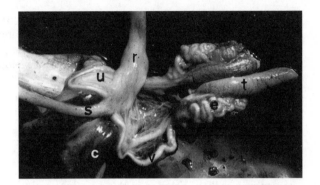

Fig. 1.4. Male genital system of *Chelodina oblonga*. *c* bulb of corpora cavernosa of penis; *e* Epididymis; *r* rectum; *s* stalk of bladder; *t* Testis; *u* ureter; *v* vas deferens

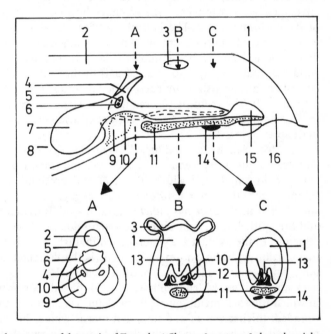

Fig. 1.5A–C. Position and structure of the penis of *Testudo*. *1* Cloaca; *2* rectum; *3* cloacal vesicle; *4* sinus urogenitalis; *5* orifice of ureter; *6* orifice of the wolffian duct; *7* vesica urinaria; *8* coelom; *9* bulbus urethralis; *10* coelomic canal; *11* corpus fibrosum; *12* corpus cavernosum; *13* seminal groove; *14* musculus retractor penis; *15* glans penis; *16* orifice of the cloaca. (After Wibault-Isebree Moens, in Bolk et al. 1933; Blüm 1986)

distended and the penis is extruded through the vent, and with full erection curves downward and slightly forward. The swollen seminal ridges close the groove into a tube that guides the semen from the openings of the vasa deferentia, which are at the base of the penis, into the female's cloaca. An additional supportive tissue, the corpus fibrosum, lies ventral to the groove and extends up to the tip of the copulatory organ, the glans penis. The glans has distinct folds, the plica externa, plica media and plica interna. The seminal groove terminates between the plicae internae. A retractor muscle assists in pulling the penis back into the cloaca after the corpora cavernosa have deflated. The penis is also laid down in females early in development, but it does not develop further and forms the clitoris.

1.3.2 Females

The paired ovaries are sheet-like organs, capable of great distension, which are symmetrically positioned in the dorsal posterior half of the coelomic cavity. The ovary comprises a stroma, nervous and vascular elements which are enveloped by a peritoneum and suspended by a mesovarium from the dorsal wall of the coelomic cavity. The juvenile ovaries are thin, elongate and sometimes lobular, spread ventrally off the kidneys, and their presumptive follicles are detectable as pale surface granules (Fig. 1.6). Mature ovaries in preovulatory condition may occupy a substantial part of the body cavity (Fig. 1.7). A follicle consists of a yellowish ovum surrounded by nurse cells and the thinly stretched ovarian wall. The ovum is set free into the coelomic cavity by a rupture of the follicular wall, a process called ovulation. Ovarian cycles will be treated in detail in Chapter 3.

The Müllerian ducts or oviducts lie ventral to the ovaries and are relatively thin and straight tubes in juveniles (Fig. 1.6). In adult chelonians the oviduct is greatly enlarged in length and width (Fig. 1.7) and produces the tertiary egg coats. Histologically, the regional differences are negligible in the oviduct of the juvenile soft-shelled turtle *Lissemys punctata*, but the adult oviduct has oviducal glands and the epithelium height differs between segments (Sen and Maiti 1990). Five segments of the oviduct are generally distinguished in chelonians: (1) the proximal portion with the ostium abdominale is the flattened and folded infundibulum, which leads to (2) the convoluted tuba uterina or glandular segment or the pars albuminifera, which is followed by (3) the isthmus or intermediate segment in which the shell membrane is formed and which forms the junction to (4) the rounded, thicker walled and unfolded uterus where the hard outer egg shell is produced, which leads to (5) the short vagina or cervix that opens, at both sides independently, ventro-laterally into the urodaeum of the cloaca.

1.4 Secondary Sexual Characteristics

In many chelonians the sexes are of almost the same size, but some groups or species show considerable sexual size dimorphism. Sexual size dimorphism will be further discussed in Chapter 4. Before and during copulation the male mounts the carapace of the female, and males of many species, in particular of the more strongly domed forms, have a concavity in the plastron which enables them to maintain the correct position on top of the female. In contrast, most females have

1.6

1.7

Fig. 1.6. Ovaries (*o*) and oviducts (*d*) of a juvenile *Erymnochelys madagascariensis* with a carapace length of 251 mm

Fig. 1.7. Ovaries and oviducts of an adult *Erymnochelys madagascariensis* with a carapace length of 395 mm during the late breeding season. *f* Vitellogenic follicles; *t* tuba uterina; *i* isthmus; *u* uterus; *v* vagina

a flat or even slightly convex plastron to provide enough space for the eggs in the body cavity (Fig. 1.8). Adult males typically also have longer tails with more distally located vents than females or juveniles. In other species, however, the plastron and the tail length differ only slightly between males and females (Fig. 1.9) and these species are often difficult to sex externally. Still, males generally show a thickening of the pre-cloacal part of the tail where the penis is contained. Adult females sometimes show scratch scars between marginal and coastal scutes from the claws of the hind feet of mating males. Males of some sea turtle species have hook-like claws on their front flippers with which, during mating, they grip the front edge of the female's carapace. Specially recurved claws on the hind feet of American box turtles of the genus *Terrapene* allow males to grip the hind edges

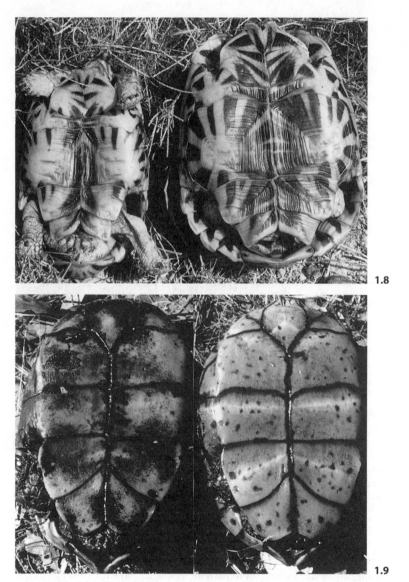

Fig. 1.8. Ventral aspect of an adult male (*left*) and an adult female (*right*) *Psammobates geometricus*

Fig. 1.9. Ventral aspect of an adult male (*left*) and an adult female (*right*) *Pseudemydura umbrina*

of the female's carapace. Many male mud turtles (*Kinosternon*) have roughened patches on the inner surface of the hind legs, which appear to have a clasping function (Pritchard 1979a). Males of some North American emydid species use exceptionally long claws on the front feet to stroke the face of females during courtship. Males of a few turtles change colour during the breeding season, which is particularly striking in *Callagur borneoensis* where skin parts as well as the shell of breeding males change in colour and brightness. In some species and populations, particularly of slider turtles, older males become melanistic (Lovich et al. 1990).

In summary, the majority of chelonians show at least some secondary sexual characteristics and dimorphism. However, besides typical males and typical females, there are often some individuals in a population which, externally, appear intermediary. Some (but not all) young males may also have rather female characteristics. A few species show externally only very minor sex differences and individuals of these species are always difficult to sex externally.

Chapter 2

Methods to Study Reproduction

The anatomical, histological, endocrinological, physiological and behavioural methods used to study reproduction of chelonians are principally the same as for other vertebrates. The box-like armour of chelonians, however, makes access to reproductive organs more difficult than in most other groups. Hurdles to reproductive biological investigations include difficulties in finding animals (or certain life stages) on a regular basis over the whole year, difficulties in readily assessing reproductive conditions in live animals (some species are even difficult to sex externally) and difficulties in obtaining blood or tissue samples. Reviews and comprehensive descriptions of general techniques for chelonian life history studies (including collecting, trapping, tracking, marking techniques, measurements and stomach flushing for dietary analyses) can be found in Gibbons (1990b), Graham (1979) and Wilbur and Morin (1988); details are beyond the scope of this book and are not presented here. Beyond the scope of this book also is the description of reproductive physiological laboratory techniques. This account deals primarily with methods to obtain reproductive biological data during field investigations.

Many valuable reproductive biological data can be collected from dead and/or preserved specimens. The collection, dissection and preservation of a large series of chelonians provided many insights into their reproductive biology. However, the sacrifice of significant proportions of a population may cause prolonged disturbances to the population dynamics, particularly in long-lived organisms like chelonians which show delayed sexual maturity. Most contemporary ecological and natural history studies of natural populations – and the ethics committees which have to approve them – do not accept such impacts. The sacrifice of large numbers of mature individuals is never justified in studies of small populations and of rare and threatened species. Knowledge of the reproductive biology of rare and threatened chelonians is still scarce, but this basic information is urgently needed to develop sound conservation strategies. In the following sections, some non-destructive techniques are described and discussed which allow the assessment of the reproductive condition of live chelonians without harm to the animals and with minimal interference with, and disturbance of, the populations under study.

2.1 Radiography

Radiography has been used for over 30 years as a non-destructive method to study oviducal eggs in live chelonians (Burbidge 1967; Gibbons and Green 1979).

Oviducal eggs can usually also be felt by palpation: a finger is pushed into the soft inguinal pocket in front of the hind legs while the female is held in an upright vertical position. If shelled eggs are present, their round shape can be felt with the finger tips. Palpation is not very accurate, particularly with small clutches in large specimens or in very small species that lay clutches of one or two eggs and have a small inguinal pocket. With radiography or X-ray photography, however, clutch size of gravid females can be determined with 100% accuracy (Gibbons 1990b). Oviducal eggs can also be measured, although a slight enlargement does occur (Graham and Petokas 1989). In addition, it is a very useful method to diagnose egg retention in the captive management of chelonians (Highfield 1996).

Radiography is limited to the detection of oviducal eggs with calcified or at least partly calcified shells (Fig. 2.1). It facilitates the assessment of neither ovarian follicles nor oviducal eggs in the early stages of gravidity during the secretion of the albumen and of the proteinaceous shell membranes. The time span during which a female's reproductive condition can be assessed by radiography is limited to a period from a few days to several weeks (only exceptionally months) per year, depending on the duration of the oviducal period and on the number of clutches, which vary between species.

Despite two decades of large-scale use of radiography in female chelonians, studies on the long-term effects of radiographs on hatchling health, fecundity and survivorship are still scarce. An early study suggested that adult turtles are less susceptible to the side effects of X-rays than either mammals or amphibians (Altland et al. 1951). Gibbons and Green (1979) reported that the hatching success of eggs from females which were X-rayed during gravidity was statistically equal to that of a control group. However, since potential long-term effects on reproductive capacities and on egg and offspring viability of individuals undergoing multiple exposures have never been investigated and remain unknown, the method should be applied with caution.

2.2 Ultra-sound Scanning

In May 1987, the survival chances of the world's most endangered chelonian species, *Pseudemydura umbrina*, received a major boost when Dr. Bruno Colombe (who, then, worked at the Department of Zoology of the University of Western Australia on the reproduction of birds) and I borrowed an ultra-sound tomographic scanner from the Department of Animal Science of the University of Western Australia and scanned various birds and reptiles at Perth Zoo, including turtles of the species *Chelodina longicollis*, *Chelodina oblonga*, *Emydura macquarii* and *Pseudemydura umbrina*. At that time, only three adult *P. umbrina* females were known to exist, none of which had layed eggs for 6 years. We found that, in turtles, ultra-sound tomographic scanning of the body cavity could be done by holding the turtles in a water bath and positioning the ultra-sound probe into the inguinal pocket cranial to the hind legs (Fig. 2.2). This allowed us to count and measure ovarian follicles, and we knew immediately that ultra-sound

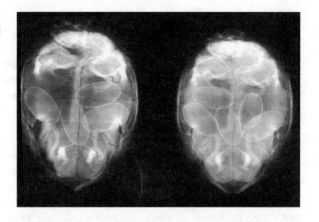

Fig. 2.1. X-ray photographs of two gravid *Pseudemydura umbrina* females with four (*left*) and five (*right*) oviducal eggs with calcified shells (Burbidge 1967, with permission)

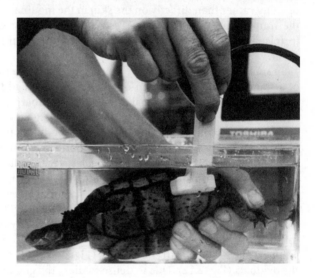

Fig. 2.2. Ultra-sound scanning of female genital organs of *Pseudemydura umbrina* in a water bath

scanning would be a major breakthrough in the study of chelonian reproduction, in particular of endangered species. Over the following months I monitored the ovarian activity of the captive *P. umbrina* females and proposed management changes to induce them to breed (Kuchling 1987, 1988a). At the same time I evaluated and defined the ultra-sound images of chelonian ovarian and oviducal structures by monitoring several *Chelodina oblonga* females throughout a breeding season (Kuchling 1989). Following the discussion of the potential of the ultra-sound method for chelonian studies in the IUCN (the World Conservation Union) Tortoise and Freshwater Turtle Specialist Group newsletter (Kuchling 1987), the method was quickly adopted by other research groups.

Ultra-sound scanning (also called ultrasonography, ultra-sound tomography, ultra-sound imaging or echography) has since been successfully used to study reproductive activity and patterns in wild as well as in captive freshwater turtles (e.g. Kuchling 1988a, 1989; Kuchling and Bradshaw 1993), terrestrial tortoises (e.g. Kuchling 1989; Robeck et al. 1990; Penninck et al. 1991; Rostal et al. 1994; Casares 1995; Gumpenberger 1996a,b) and marine turtles (e.g. Rostal et al. 1990, 1996; Plotkin et al. 1997). The corneal shields and bone plates of the chelonian shell block ultra-sound waves, restricting the "acoustic window" for scanning the body cavity to the soft skin of the inguinal area. Large tortoises and turtles are best scanned by using water-soluble coupling gel as a transmission medium between the skin and the probe, whereas many smaller species are best scanned in a water bath with the water as transmission medium (Fig. 2.2). Manual restraint of the animals or, in giant tortoises, soliciting their cooperation by scratching their neck and limbs (Casares 1995) is generally sufficient and less stressful than restraining devices which have also been used for tortoise ultra-sound examinations (Robeck et al. 1990). Scanning in a water bath is particularly useful in small species with a narrow opening between the carapace and plastron in which the probe may not fit, since the probe can be positioned in front of the shell opening rather than directly on the skin in the inguinal pocket – without reducing the transmission quality. In terrestrial species, only the posterior part of the tortoise is tilted into a shallow water bath until the inguinal pockets are under water.

In small terrestrial species, an alternative technique to the water bath is the use of a stand-off pad in the form of a coupling-gel-filled finger of an examination glove inserted into the inguinal pocket (Gumpenberger 1996b). In small, highly-domed species with a narrow shell opening (e.g. species of the South African genus *Psammobates*), scanning of the entire body cavity is not possible with most ultra-sound probes; only a section of the body cavity and the reproductive organs can be visualised (Fig. 2.3). In these species, very small probe heads which fit into the narrow shell opening would give much better results.

A major advantage of ultra-sound scanning over radiography is that the former allows the assessment of not only shelled oviducal eggs (Figs. 2.3, 2.4), which are present during a limited period of gravidity only, but also vitellogenic follicles with a diameter larger than 2–4 mm (depending on the quality of the scanner; Figs. 2.5–2.8), atretic follicles (Figs. 2.9–2.11) and freshly ovulated ova in the oviducts (Fig. 2.12). Due to dense folding, the glandular segments of oviducts (or tuba uterina) may appear as multilayered tubes during ultra-sound examinations (Fig. 2.13). Vitellogenesis in chelonians takes many months and an ultra-sound examination allows collection of data on the female reproductive condition more or less at any time of the year. The occurrence of batches of different size classes of follicles indicates the possible number of clutches per reproductive season (Moll 1979). All modern ultra-sound devices have electronic calliper systems which allow one to measure distances on the screen, e.g. diameters of follicles, to an accuracy of about 1 mm.

The limitations of ultra-sound scanning include difficulties in readily identifying corpora lutea and male reproductive organs. In contrast to egg counting on X-ray photographs, counting of follicles and eggs by ultra-sound scanning is never 100% accurate, because examinations of the body cavity from two sides have to be summed and some items may remain undetectable in the shadow of closer structures. A quantitative assessment is impossible in large chelonians (giant tortoises, sea turtles) in which only the caudal-most aspects of the ovaries or oviducts can

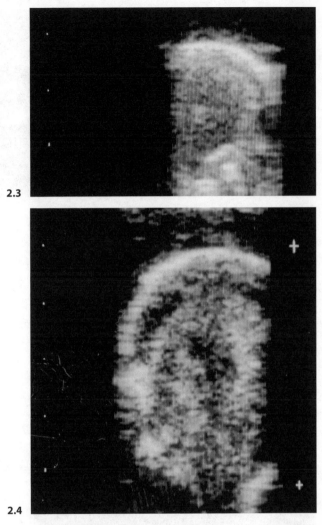

2.3

2.4

Fig. 2.3. *Psammobates geometricus*. Ultra-sound tomographic picture of a section of a shelled, calcified egg in the oviduct; the scale (*white dots*) represents centimetres

Fig. 2.4. *Geochelone yniphora*. Ultra-sound tomographic picture of a section of a shelled, calcified oviducal egg; callipers (*white crosses*) indicate 42 mm, the scale (*white dots*) represents centimetres

be visualised. A quantitative assessment of follicles and eggs by ultra-sound scanning is only feasible in turtles of intermediate size (about 300–3000 g) and small or intermediate egg numbers per clutch (maybe up to 20). Furthermore, best results are obtained in species with relatively large shell openings, e.g. a 92% accuracy in counting follicles and eggs was found in *Chelodina oblonga* which has a clutch size of 8–16 eggs (Kuchling 1989).

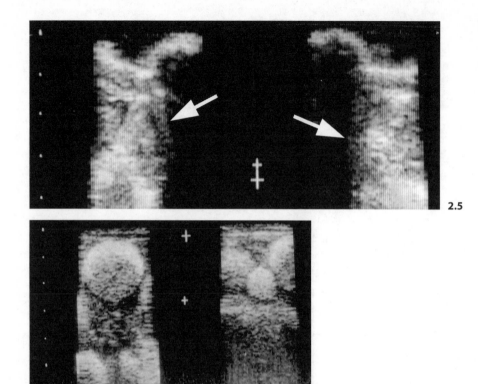

2.5

2.6

2.7

Fig. 2.5. *Pyxis arachnoides*. Ultra-sound tomographic picture of ovarian follicles with diameters of 3–4 mm (*arrows*) at start of vitellogenesis; callipers (*white crosses*) indicate 4 mm, the scale (*white dots*) represents centimetres

Fig. 2.6. *Geochelone yniphora*. Ultra-sound tomographic picture of vitellogenic follicles with diameters of 12 and 23 mm; callipers (*white crosses*) indicate 23 mm, the scale (*white dots*) represents centimetres

Fig. 2.7. *Geochelone gigantea*. Ultra-sound tomographic picture of vitellogenic follicles with diameters of 25 and 30 mm; callipers (*white crosses*) indicate 30 mm, the scale (*white dots*) represents centimetres

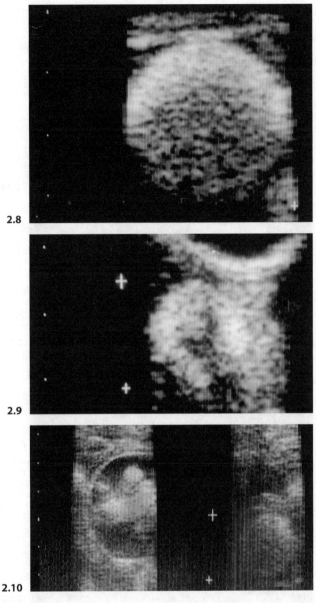

2.8

2.9

2.10

Fig. 2.8. *Geochelone yniphora*. Ultra-sound tomographic picture of a preovulatory follicle with a diameter of 30 mm; the scale (*white dots*) represents centimetres

Fig. 2.9. *Pelusios castanoides*. Ultra-sound tomographic picture of a follicle with a diameter of 16 mm in early atresia. Note slight deviations from spherical shape and area with lower echodensity; callipers (*white crosses*) indicate 16 mm, the scale (*white dots*) represents centimetres

Fig. 2.10. *Geochelone nigra*. Ultrasound tomographic picture of a section of an atretic follicle with a diameter of 40 mm (*left*) and of a vitellogenic follicle with a diameter of 23 mm (*right*). Note areas with varying degrees of echodensity in atretic follicle, indicating degeneration of yolk globules and progressive removal of yolk constituents; callipers (*white crosses*) indicate 23 mm, the scale (*white dots*) represents centimetres

23

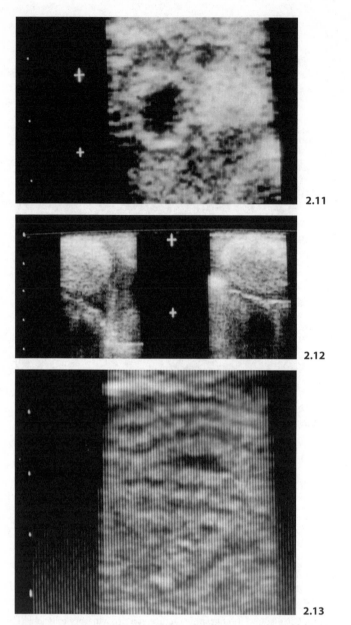

Fig. 2.11. *Geochelone yniphora.* Ultra-sound tomographic picture of a follicle in late atresia. Anechoic centre indicates removal of most of the yolk material; callipers (*white crosses*) indicate 12 mm, the scale (*white dots*) represents centimetres

Fig. 2.12. *Geochelone radiata.* Ultra-sound tomographic picture of freshly ovulated ova in tuba uterina (longitudinal section). Note elongate appearance of ova and echodense tuba uterina; callipers (*white crosses*) indicate 25 mm, the scale (*white dots*) represents centimetres

Fig. 2.13. *Geochelone nigra.* Ultra-sound tomographic picture of tuba uterina (cross section); the scale (*white dots*) represents centimetres

24

2.3 Laparoscopy and Biopsies

Laparoscopy is a surgical procedure which allows the visual examination of organs in the coelomic cavity through a telescope which incorporates fibre optics light transmission from a cold light source. A rigid endoscope of about 4 mm diameter is generally sufficient. I perform laparoscopy on turtles under general anaesthesia (ketamine–HCL applied intramuscularly at a dosage of 60–80 mg/kg body weight: see Kuchling 1989; see Schildger et al. 1993 for dosages required in various chelonian species), although local anaesthesia is considered adequate for laparoscopic procedures in various reptiles (e.g. the Tuatara *Sphenodon punctatus*: Cree et al. 1991), marine turtles (Limpus and Reed 1985) and giant tortoises (Robeck et al. 1990). A small skin incision is made in the inguinal pocket posterior of the bony bridge, and muscles and serosa are bluntly perforated in craniolateral direction with a trochar and cannula. Air has to be insufflated into the body cavity to a pressure of about 10 mm Hg to allow for sufficient lens-to-organ distance and to enhance focusing ability. After completion of the examination the air has to be voided from the body cavity by gently pushing the hind limbs into the inguinal pockets (particularly important in aquatic species) and the incision is closed with sutures of synthetic absorbable material (Vicryl).

Laparoscopic examination of reproductive organs allows a qualitative assessment of the reproductive condition. In males, colour and size of testes give an indication of spermatogenetic activity, and turgescent, white sperm ducts of the epididymis (Fig. 2.14) indicate stored sperm. In ovaries, previtellogenic follicles appear as small white spots or discs; vitellogenic follicles are bright yellow spheres with clear red blood vessels; corpora lutea are whitish tissue knobs with an opening or slit through which the ovum has emerged; atretic follicles are dark yellow to orange with diffuse blood vessels, and often are not perfectly spherical; and oviducal eggs appear as oval inclusions in the whitish uterine tubes.

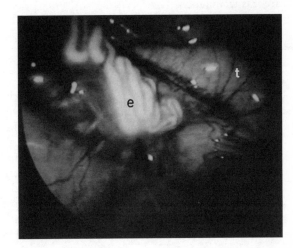

Fig. 2.14. *Chelodina oblonga.* Endoscopic picture of testis (*t*) and epididymis (*e*)

Laparoscopy can be combined with the sampling of tissue biopsies from internal organs. Some endoscope trochars include cannulae for biopsy instruments, but the small models do not, and, for the insertion of the tool, it is necessary to provide a second entry point into the coelomic cavity. Biopsies are particularly useful in studies of male reproduction, because the only method to assess properly spermatogenetic activity in chelonians is the histological examination of the seminiferous tubules of the testis. Biopsy needles are not useful to take testis biopsies in chelonians, because their testes hang like pendules from the dorsal coelomic wall, suspended by the mesorchium. The testes tend to slip away during attempts to push biopsy needles through the tunica albuginea. Testis biopsies are better taken with a biopsy punch forceps with cutting edges, which grasps a part of the testis and cuts out a small tissue sample. To avoid haemorrhage after biopsies it is important to select testis sections without large blood vessels in the tunica albuginea.

The limitations of laparoscopy include the inability to measure structures accurately: their size can only be estimated, which is difficult through wide-angle lenses. The accuracy of size estimates can be improved if reference objects of known dimensions, e.g. a biopsy forceps, are held adjacent to a structure. This, however, requires a second entry point into the coelomic cavity. Due to the congestions caused by enlarged ovaries and convoluted oviducts in the body cavity of females in breeding condition, laparoscopy generally does not allow one to count numbers of follicles, corpora lutea or oviducal eggs.

A technique which provides results similar to laparoscopy is soft tissue laparotomy of chelonians, but it requires more extensive surgery and anaesthesia. A horizontal skin incision is made in the inguinal pocket immediately cranial to the hind limb. The coelomic viscera can then be inspected with the aid of an otoscope speculum (Brannian 1984). This method is particularly suitable if operations have to be performed (e.g. salpingotomy and surgical removal of eggs). Laparotomy has also been described as a method to sex immature sea turtles (Owens 1995), but if the scope is only inspection of the gonads and sampling of biopsies, laparoscopy is the preferred technique because the surgical procedures are less traumatic.

2.4 Blood Sampling

Parameters to assess reproductive conditions which can be measured in blood or plasma samples include hormones which regulate reproductive functions and processes, as well as substances, which are to be incorporated into the eggs such as vitellogenin, lipids and calcium. Jacobson et al. (1992) reviewed published blood sampling techniques for chelonians and concluded that the best blood samples are obtained by inserting a butterfly catheter or a needle into a blood vessel and by careful aspiration of blood into a syringe. According to these authors, potassium EDTA caused haemolysis of turtle red blood cells and they recommended the use of lithium heparin as an anticoagulant if blood is collected for haematological and biochemical studies. Cardiac punctures

(Stephens and Creekmore 1983) and orbital sinus sampling (Nagy and Medica 1986), which have also been used in chelonian studies, may dilute the blood sample with extravascular fluids and are not recommended for haematological studies.

The trick is to find a suitable blood vessel, which, depending upon the species, may be difficult. In desert tortoises *Gopherus agassizii*, Jacobson et al. (1992) sampled blood from the jugular vein and found that the jugular vein and the carotid artery were the only peripheral blood vessels that could be consistently visualised. I found the jugular vein very suitable for blood sampling in all Australian chelid turtles except *Pseudemydura umbrina* in which, in most cases, it cannot be visualised. Sampling blood from the jugular vein requires cephalic restraint and extension of the neck, which is often easier in pleurodire (side-necked) than cryptodire (hidden-necked) turtles.

A common site for blood sampling in chelonians is the forelimb using the scapular vein, brachial vein and brachial artery (Avery and Vitt 1984). An axillary sinus lies superficially and medially to the large tendon which can be palpated. A foreleg is restrained in extended position and the needle inserted at the inside of the axillary area. In most cases these vessels cannot be visualised, and obtaining blood requires blind probing with the needle to puncture a vessel or the sinus. Sampling blood from the foreleg often results in haemodilution with lymph and may, therefore, not be truly representative of blood (Jacobson et al. 1992).

The sample techniques above work better in medium-sized and large species than in small ones. In most chelonians, including many small ones, it is also possible to sample blood from either the dorsal coccygeal vein, which runs along the midline of the dorsal aspect of the coccygeal vertebrae, or from a blood sinus in the tail. The needle is carefully inserted from dorsally into the midline of the tail between two vertebrae until blood can be aspired. This is, for example, the best blood sampling technique in *Pseudemydura umbrina*, Australia's smallest and one of the world's most endangered chelonians. It is, however, not a practical technique in small species with small tails which can be deeply retracted into a narrow shell opening, for example in the small South African tortoise species.

A further technique which is useful in small species is to sample blood from a blood sinus which is located dorsally of the 7th cervical vertebra (P. King, pers. comm.). This method has been successfully used in haemoglobin studies of *Kinosternon odoratum*, *Clemmys guttata* and *Terrapene carolina* (King and Heatwole 1997). A heparinised needle is inserted from anteriorly just under the carapace along its midline, dorsally of the neck, until it reaches the point where the first trunk vertebra is fused with the shell. Blood is then aspired. This method does not require the extension or restraint of any body part and the animal can remain fully retracted in its shell, which reduces stress during the sampling procedure.

In sea turtles, blood is best sampled from the bilateral dorsal cervical sinus. The head is pulled gently forward and downward until the neck is fully outstretched and the needle entry to the cervical sinus should be perpendicular to the skin and just lateral to the midline. This method works in larger

27

specimens of all species (Owens and Ruiz 1980) and in hatchling loggerhead turtles, *Caretta caretta* (Bennett 1986), but may be difficult in hatchlings of other species.

2.5 Considerations Regarding Captive Maintenance

Most specimens in studies of captive maintained chelonians will have been collected in the wild. Only a few North American and Chinese species are bred in large-scale commercial farms and sold in the pet trade. Sea turtle breeding or ranching is limited and controversial. Some zoos and private breeders are successfully breeding a few chelonians, but, despite these potential sources, few reproductive data have been obtained from chelonians that were produced in captivity. One reason for this phenomenon is the chelonian trait of delayed maturity, which means that captive bred animals have to be raised for many years or over a decade before they mature, something few researchers can afford.

In studies of captive chelonians it is important to consider and to provide details of where the animals originated from and how long they were maintained in captivity before the start of the study. In the live animal trade, many chelonians are able to survive incredibly harsh and stressful conditions and treatments. Such conditions, however, may dramatically alter reproductive processes. Male *Chelydra serpentina*, caught during spring and early summer by a commercial trader, maintained regressed testes in captivity, females did not ovulate and follicular growth slowed down dramatically (Mahmoud and Licht 1997). Only 4 out of 130 *Kinosternon odoratum* females which were commercially acquired during autumn, despite having active ovaries, ovulated in the laboratory during the following reproductive season (Mendonça 1987a). However, 6 out of 7 *C. serpentina*, after transfer into captivity in autumn, and without going through the commercial trade, ovulated in the following spring (Mahmoud and Licht 1997). If animals spend prolonged periods in the trade it may take many years until they regain normal reproductive functions, although their general condition may seem satisfactory.

Even if much care is taken to ensure that freshly collected chelonians are transferred into captivity without excessive stress, the transfer into captivity and/ or the captive conditions may still inhibit or alter reproductive processes. In *Pseudemydura umbrina*, for example, the careful, quick (1–2 h) and short (over a distance of 20 or 30 km) transfer of females from the wild habitat into large outdoor enclosures with grossly similar climatic and environmental conditions inhibited ovulation and/or altered their reproductive processes for a period of 1–3 years (Kuchling and Bradshaw 1993). There are species-specific differences in the sensitivity to disturbances. In some species ovulation seems to be less sensitive to stress effects of transfer into captivity and females with preovulatory follicles may ovulate in the laboratory, for example *Chrysemys picta* (Ganzhorn and Licht 1983) or *Chelodina oblonga* (G. Kuchling, unpubl. data). The conditions of maintenance obviously also play a role: preovulatory *C. serpentina* females ovulated after being taken into a university holding facility, but did not ovulate in the holding facilities of a commercial trader (Mahmoud and Licht 1997).

Androgen levels and spermatogenesis were altered in captive male *Testudo hermanni* compared with the wild population from which they originated, even though the tortoises were kept in outdoor enclosures (Kuchling 1981; Kuchling et al. 1981). Captivity has also been shown to depress plasma steroid levels in North American freshwater turtles (Mendonça 1987a; Mahmoud and Licht 1997). Even long-term, well established captive turtles may react to rather slight management changes by suppressing reproduction. For example, three long-term captive *Chelus fimbriatus*, after reaching maturity and regularly breeding in their tank, stopped reproducing (mating as well as egg production) for 2 years after the lighting intensity above their tank was temporarily increased for filming purposes (Schaefer 1986).

It is, therefore, important to consider and to state clearly if data were procured from wild or captive animals, and under what conditions and over what period animals were maintained in captivity. The sensitivity of the reproductive processes of many chelonians to stressful conditions seriously limits the value of many laboratory experiments. In experimental setups it is imperative to include a control group which is housed under similar conditions as the experimental group and not, for example, in spacious outdoor enclosures. This may seem to be self evident, but disregard for this basic requirement diminishes the value of several experimental studies (see Sect. 6.1.1).

To summarise, captivity, the conditions of captive maintenance and temporary changes of the management routine may have dramatic implications for the short-term as well as long-term reproductive performance of chelonians. If aspects of chelonian reproduction are studied in captive specimens, it is clearly important to consider the species-specific requirements for an adequate captive environment. It is beyond the scope of this book to treat this topic. The literature on maintenance techniques, health care and general captive requirements of the various chelonian groups is extensive. Recent reviews are, for example, provided by Highfield (1996) and Rogner (1995, 1996).

Chapter 3

Gonadal Cycles and Gamete Production

According to the germinal line theory (Weismann 1885), metazoan organisms can be reduced to two components, the soma and the germinal line. The soma constitutes the bulk of the body, has a limited life span and is the part which dies; the germinal line is potentially immortal. It is represented by the primordial sex cells and by the male and female gametes which arise from them and link the successive generations of a species into a continuous lineage. The germinal line is often recognisable early in embryological development. The primordial germ cells of *Chrysemys picta* arise in a posteriorly directed horseshoe-shaped zone of the extra embryonic hypoblast (endoderm); they are distinguishable from the surrounding endodermal cells by their large size, and the fact that they contain yolk granules; they migrate interstitially, by active, independent, amoeboidal movement to the embryonic anlagen of the gonads, the germinal ridges (Allen 1906). Once the primordial germ cells of chelonians are grouped in the hypoblastic crescent that surrounds the caudal extremity of the embryo, they reach the germinal ridges by amoeboid movements through the splanchnopleure and lining of the digestive tract (Hubert 1985). The primordial germ cells, or the oogonia and spermatogonia arising from their mitotic divisions, remain dormant, often for long periods, until the organism reaches reproductive maturity. Only then do they undergo meiosis to become gametes – the eggs and spermatozoa.

The major function of gonads is the production of gametes; the second important function is the secretion of hormones which determine the development of sexual organs and sexual dimorphism, influence behaviour, prepare the sexual organs to harbour the gametes and facilitate fertilisation and the secretion of the outer layers of the egg. In most if not all chelonians, periods of reproductive activity alternate with periods of inactivity. The relationship of these cycles to season, climate and other environmental parameters is discussed in Chapter 5; gonadal hormone secretion and the regulatory and control mechanisms are treated in Chapters 6 and 7. The emphasis of this chapter is to describe the processes of gamete production and the changes in the reproductive organs which are the basis of, and coupled with, the different phases of the reproductive cycle.

3.1 Sexual Maturity

Chelonians, when compared with many other reptiles of similar size, are characterised by extended longevity and delayed maturity. Their juvenile and subadult phase lasts several years and, in some of the largest forms, even decades. Onset of sexual maturity seems to depend on the attainment of a certain size as well as on age (Moll 1979), although some investigations suggest that, in given populations, age may be more important in controlling sexual maturity (Vogt 1997b). Even in one species there may be considerable variability, within and among populations, of the size and age at which different individuals reach maturity (see Sect. 9.1.1).

Sexual maturity is reached in males with the first production of spermatozoa. In dissected animals this can be ascertained by the appearance of the testis which enlarges during seasonal spermatogenesis and/or the appearance of the epididymis in which, at the time of testicular regression, the presence of sperm is indicated by the engorgement and pallidity of the convoluted ductus epididymidis. In most population studies sexual maturity in males is determined by the expression of secondary sexual characteristics such as elongation and thickening of the preanal portion of the tail. Since gonadal hormones influence these sexual characteristics, their levels in the blood may be a reasonable measure of maturity in males. In males as well as females, growth declines with the onset of puberty or maturity, as evidenced by a narrowing of the annual growth rings of epidermal scutes. In females, however, external characteristics on their own do not facilitate an accurate identification of maturity. Sexual maturity of females is defined as the capacity for producing eggs during the next breeding season (Gibbons and Green 1990). Cycles of vitellogenic oocyte growth and, therefore, hormonal changes may start a few years before maturity is reached: *Pseudemydura umbrina* ovaries cycle at least 2 years before ovulation occurs, with follicles reaching preovulatory size and then being reabsorbed (Fig. 3.1). Many female chelonians may have a subadult stage of several years, during which vitellogenetic cycles occur, before reaching full maturity with the onset of egg production. Reliable expressions of maturity in females are the occurrence of oviductal eggs and/or corpora lutea or ovulation scars at the ovary.

3.2 Ovarian Cycle and Egg Production

Oogenesis, the formation of oocytes from oogonia, occurs throughout the reproductive life of reptiles, in contrast to Petromyzontia, Elasmobranchii, a few Teleostei, Aves and Mammalia, in which oogenesis is restricted to the embryonic phase (Blüm 1986). The oogonia divide mitotically and eventually form oocytes which then are surrounded by follicle cells and form primordial follicles. The quantitative aspects of this process, the number of mitotic divisions of oogonia during reproductive cycles, are not known for any reptilian species (Guraya 1989). Vitellogenesis, the accumulation of yolk material in the oocyte, are one of the most conspicuous features of the reptilian ovarian cycle, and is generally used to define seasonal patterns in ovarian activity. Maturation and ovulation is little studied in reptiles, generally, and in chelonians, particularly. The oviducal period of eggs and the luteal phase precede oviposition. Follicular atresia may occur at all

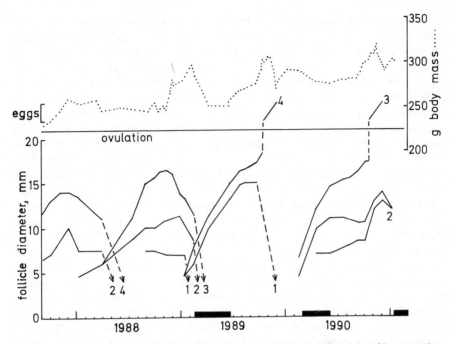

Fig. 3.1. *Pseudemydura umbrina.* Annual ovarian cycles of a subadult female reaching maturity, showing ovarian follicle growth; ovulation (*horizontal line*); oviposition; atresia (*solid and dashed lines*); female body mass; aestivation (*black bars*); numbers of eggs laid or numbers of follicles which disappeared following atresia (*numbers*)

stages of follicular development and, in chelonians, most commonly occurs at the end of the nesting season. The degree of ovarian regression between consecutive cycles is highly variable between chelonian populations and species.

3.2.1 Oogenesis and Folliculogenesis

The sites for the multiplication of oogonia and for the formation of primordial oocytes are the germinal beds or cell nests of the ovary. In *Trachemys scripta*, oogonia or preoogonia are mostly at the surface of the ovary in direct contact with the peritoneal cavity, and not below the tunica albuginea (Callebaut et al. 1997). Oogonia proliferate through mitosis and form groups of oogonia. Oogonial proliferation in chelonians seems to be a seasonal event. In the box turtle *Terrapene carolina*, a peak of mitotic activity of oogonia was observed during July and August following the period of ovulation and egg laying (Altland 1951). I do not know of any other similar study of seasonal changes of the number of oogonia and oogonial mitoses in chelonians. There follows a growth phase of the oogonia which become partially surrounded and separated from each other by prefollicular cells, which is a transitional stage from oogonia to first-order oocytes. In *T. scripta*, all oocytes with a diameter of 200 µm or more are surrounded by voluminous lacunae and are enclosed in a follicle with a superficial crater, bulging at the surface of the ovary (Callebaut et al. 1997). The first-order oocytes then enter the meiotic prophase. Risley (1933a) noted that oogonia in the ovaries of embryos of *Kinosternon odoratum* have already entered the meiotic prophase, but this may be an exception. Oogenesis continues throughout the reproductive life of chelonians. The oocytes keep growing and differentiate by storing in their cytoplasm the reserves necessary for the development of the embryo. This process of vitellogenesis is mediated by cells which form the follicle of the egg. The formation of the follicle (folliculogenesis), therefore, is a prerequisite for the occurrence of vitellogenesis. The follicle cells probably stem from the coelomic epithelium and migrate, often as pregranulosa cells, into the germinal beds where they surround individual oocytes (Blüm 1986). During folliculogenesis the oocyte shifts from the germinal bed towards the stroma of the ovarian medulla (Guraya 1989).

Guraya (1989) demonstrated that the growing previtellogenic oocyte of the soft-shelled turtle *Lissemys punctata* shows an extensive development of nucleoli inside the nuclear envelope (germinal vesicle), which he attributed to the amplification of ribosomal genes (rDNA). This process in chelonians seems to be similar to that in fish and amphibians. It differs from other reptilian (squamate) and avian oocytes which show a relatively lesser development of nucleoli in the germinal vesicle: in these groups the amplification of ribosomal genes appears to occur in the follicle cell nuclei. This pattern in chelonians reflects their phylogenetic antiquity, and their close relationship to the earliest reptiles which evolved from amphibian ancestors.

The follicular wall which forms around the developing and maturing oocyte consists of the zona pelludica, follicular epithelium, basal lamina and thecal layers which show changes during the growth of the oocyte. The follicle cells of developing previtellogenic oocytes are actively involved in the synthesis of proteins, glycogen, phospholipids, RNA and ribosomes, which appear to be transported into the ooplasm. In contrast to lizards and snakes, which have a polymorphic follicular epithelium that connects to the growing previtellogenic oocyte by intercellular bridges, chelonians (and birds) show follicular cells of more or less similar morphology (cuboidal or columnar) without intercellular bridges. The granulosa of the developing follicle usually consists of one layer of cells (Hubert 1985). In *T. scripta*, however, granulosa cells laterally – and in the immediate neighbourhood of the germinal vesicle – form a two-nuclear-thick layer (Callebaut et al. 1997). The follicular epithelium of chelonians shows light and dark cells. The latter appear to be degenerating cells, the products of which are transported to the developing oocyte for its nutritional and developmental requirements. In the developing follicles of chelonians the plasma membrane of the follicle cells forms annular-shaped organelles called "transosomes" or "lining bodies" which are released into the perivitelline space and finally engulfed by the oocyte. Between the oocyte surface and the follicular epithelium is the acellular zona pelludica (or vitelline membrane) which consists of carbohydrates and proteins. During the early stages of follicular development the zona pelludica is an homogeneous layer, but with further growth of the follicle it differentiates into an outer homogeneous layer and an inner striated layer. In vitellogenic oocytes, the striated layer increases in width and becomes three times thicker than that of previtellogenic oocytes (Guraya 1989).

Three successive developmental stages of intrafollicular oocytes can be distinguished in *T. scripta*: (1) the prelampbrush stage with chromosomes in diplotene; (2) the lampbrush stage which begins when the oocyte has a diameter of 600–700 μm, with chromosomes presenting the lampbrush chromosome configuration with lateral loops; (3) the postlampbrush stage which starts when the oocyte has a diameter of 3–4 mm. At this stage the germinal vesicle penetrates the deeper part of the cortex, and a germinal disc forms in which subcortical ooplasmic organelles develop and where the beginning of the assemblage of small yolk spheres takes place. In the 10-mm-diameter oocyte, the germinal vesicle has penetrated through the cortex. At its peak development, just before maturation, the germinal vesicle is flattened at the surface of the ooplasm, but still has a yolk cap (Callebaut et al. 1997).

3.2.2 Vitellogenesis

Vitellogenesis is the process of hepatic synthesis and secretion of yolk protein, its transport via the blood stream and subsequent deposition in the oocytes. The liver synthesises yolk in the form of vitellogenin, a lipoglycophosphoprotein of high molecular weight, under the influence of oestrogen produced by the wall of maturing follicles (see Chap. 7). After the various ooplasmic organelles have multiplied and accumulated during the growth of previtellogenic oocytes, the deposition of yolk starts. The capillaries of the theca deliver the plasma vitellogenin to the follicle cells where it is selectively taken up, transported into the oocyte and deposited in the ooplasm in the form of complex, discrete bodies.

Chrysemys picta vitellogenin is composed of two polypeptides of equal size with molecular weights of about 210–220 kDa which is similar to the molecular weight of chicken (bird) vitellogenin and heavier than *Xenopus* (frog) vitellogenin (195 kDa); however, the amino acid composition of *Chrysemys* vitellogenin is strikingly similar to that of *Xenopus*, and both these vitellogenins show little degradation during precipitation from the plasma. In contrast, chicken vitellogenin, under the same conditions, degrades into smaller polypeptides (Ho et al. 1980). A specific antiserum against *Chrysemys* (Emydidae) vitellogenin showed cross-reactivity with vitellogenin in sera of several other chelonians – *Clemmys guttata*, *Clemmys marmorata* (Emydidae); *Kinosternon minor*, *Kinosternon odoratum* (Kinosternidae); *Chelydra serpentina* (Chelydridae); *Chelonia mydas* (Cheloniidae) – but showed no cross-reactivity with sera from Squamata and Crocodilia (Gapp et al. 1979).

After vitellogenin has been transported into the oocyte it is rapidly converted into two different proteins, lipovitellin and phosvitin. These are then recombined to form a crystalline structure. *Chrysemys* yolk has a different amino acid composition from *Chrysemys* vitellogenin; it has a relatively simple composition of one phosvitin with a molecular weight of 44 kDa and one lipovitellin composed of two major subunits with 120 and 85 kDa; the high serine content (50% of the total amino acids) of *Chrysemys* phosvitin is comparable with a high serine content (56%) in *Gallus* and *Xenopus* phosvitin (Ho et al. 1980). These similarities of the *Xenopus*, *Chrysemys* and *Gallus* vitellogenins and phosvitins suggest evolutionary conservatism regarding yolk proteins and their mechanisms of synthesis. Surprisingly, Mulaa and Aboderin (1992) reported that yolk proteins in oocytes of *Kinixys erosa* (Testudinidae) are composed of two phosphoglycoproteins (phosvitins) with molecular weights of 68 084 and 29 599 Da with relatively lower serine contents of 45% and 16.5%, respectively. This study of *Kinixys erosa* yolk proteins suggests the possibility of pronounced differences in yolk between different chelonians, but yolk analyses of more species are needed with standardised techniques before this question can be answered.

Two types of yolk bodies are formed in the oocyte, fatty yolk and proteid yolk globules. After the various organelles such as the yolk nucleus substance, endoplasmatic reticulum, ribosomes, mitochondria and Golgi bodies have accumulated in the cortical ooplasm of the growing oocyte, the deposition of proteid yolk bodies starts in the central ooplasm, where only a sparse distribution of organelles is seen; initially, yolk vesicles or vacuoles are developed in the central ooplasm. As the vitellogenic follicles grow, protein and carbohydrates accumulate in the yolk vesicles or vacuoles (Guraya 1989). In *T. scripta*, yolk globules increase in volume by fusion. Even in the largest oocytes, only round yolk spheres are assembled and no polyhedric yolk units or platelets are formed. The successive yolk layers form concentric spheres around the centre of the oocyte. No structure occurs comparable to the avian nucleus of Pander, a large cushion-like mass of white yolk below the germinal disc (Callebaut et al. 1997).

The oocyte can take up various other substances with high molecular weight in addition to vitellogenin, for example lipids. An apolipoprotein B-100 with a molecular weight of approximately 350 kDa, which has a role in lipid transport and may be essential for cellular uptake of lipids, has been isolated from the plasma of reproductively active female *Chrysemys picta* and seems to have a high degree of structural homologies with chicken apolipoprotein B-100, but not with mammalian apolipoprotein B-100 (Perez et al. 1992). Fatty yolk in the oocytes of turtles consists of neutral fat or triglyceride globules of variable size. The total lipid proportions of the egg yolk vary between species: it is 29.8% in *Chrysemys picta* and about 14% in *Chelydra serpentina* and *Emydoidea blandingii*; in all those species >63% of the total lipids occurs as triacylglycerol, an energy storage form (Rowe et al. 1995). In contrast to lizards and snakes, which have discrete abdominal fat bodies (corpora adiposa) in close proximity to the gonads that supply lipids for vitellogenesis to the gonads, chelonians do not show localised abdominal fat bodies; instead, many small fat deposits ("pads") are dispersed throughout the abdominal cavity from which lipids are mobilised during vitellogenesis. Chaikoff and Entenman (1946) described higher plasma levels of cholesterol and fatty acids in female turtles with active ovaries than in females with inactive ovaries. A decline in carcass lipids was found to be associated with an increase in ovarian mass in *Kinosternon odoratum*, and seasonal peaks of plasma lipids were temporally associated with ovarian development and carcass fat storage (McPherson and Marion 1982). The liver seems to be the organ of both manufacture and utilisation of lipids, which are then transported via the blood into peripheral tissues where they are stored or utilised. The mechanisms of transport of lipid yolk precursors from the egg envelopes into the oocytes are not known, but pinocytosis and phagocytosis may play some significant role (Guraya 1989). During the final growth period of the oocyte of *T. scripta*, highly osmiophilic, alcohol-insoluble satellite yolk or egg oil accumulates between the protein yolk globules. This yellow-stained satellite yolk seems to contribute to the yellow colour of the large oocytes (Callebaut et al. 1997).

In most birds, crocodiles and squamates, individual follicles accumulate yolk and grow to preovulatory size in time spans of days or weeks. The growth of individual follicles of chelonians typically takes many months, often two-thirds of a year, and sometimes even several years. With regard to the time needed for vitellogenesis by the various reptile groups, only the Tuatara of New Zealand (genus *Sphenodon*, order Rhynchocephalia) is comparable to the chelonians with the longest cycles: the Tuatara lays eggs every 4 years and vitellogenesis of individual follicles is spread over 3 years (Cree et al. 1992).

In many temperate-zone chelonians and some subtropical ones – more data are available for these than for tropical species – vitellogenesis starts in late summer or autumn and continues until spring when ovulation occurs (see Chap. 5). In climates with cold winters, where vitellogenesis ceases during hibernation, completion of follicular enlargement for the first clutch may occur before or after hibernation. In most chelonians it is possible to distinguish groups of growing follicles of several sizes. In turtles laying multiple clutches these batches of follicles of different sizes represent subsequent clutches (Moll 1979), but turtles which, typically, lay only one clutch per year may also have more than one size class of growing follicles in their ovaries. There is little coherent information in the literature about the fate of the batches of smaller vitellogenic follicles which do not ovulate in a given year; whether they persist until the next period of vitellogenesis, complete growth and ovulate in the following breeding season or whether they become atretic and are reabsorbed is not known. Both patterns may occur, depending upon species and populations, but neither persistence nor atresia are well studied in wild populations. It seems that in several cryptodire species of the temperate northern hemisphere, at least, some follicles may start vitellogenesis 2 or 3 years before reaching preovulatory size, but most papers on vitellogenetic cycles of chelonians do not present or discuss the situation of the smaller classes of vitellogenic follicles, possibly because they are found all year round. In a comprehensive study, Callard et al. (1978) inferred from the dissection of over a hundred *Chrysemys picta* females from Wisconsin that individual vitellogenic follicles take 3 years to grow to preovulatory size, although females produce eggs on an annual basis. A similar pattern of extended vitellogenesis, but annual oviposition, may occur in other Emydidae – as well as in the Bataguridae *Mauremys leprosa* (Combescot 1955a); in the Kinosternidae *Kinosternon odoratum* (McPherson and Marion 1981a) and *Kinosternon flavescens*

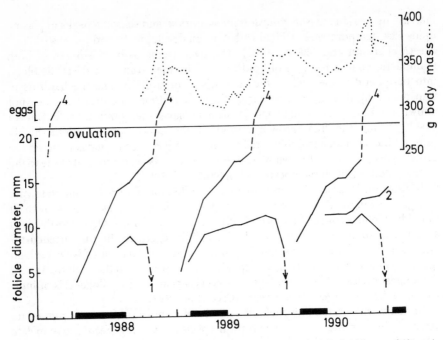

Fig. 3.2. *Pseudemydura umbrina.* Annual ovarian cycles of an adult female; *symbols* as in Fig. 3.1

(Christiansen and Dunham 1972); and in the Chelydridae *Chelydra serpentina* (White and Murphy 1973).

In the Chelidae of temperate Australia, vitellogenesis of individual follicles is typically completed in 1 year and normally does not seem to extend over several years. The application of ultra-sound scanning has allowed monitoring of follicular growth as well as ovulation and follicular atresia in individual turtles over several years (Kuchling 1989; Kuchling and Bradshaw 1993), and facilitated an assessment of the growth pattern of oocytes. Figure 3.2 shows the follicular cycles of a *Pseudemydura umbrina* female which has been monitored over several years. Once follicles reach the size of 3–4 mm diameter (Fig. 3.3), typically by February or March during the southern hemisphere summer, most keep growing (Fig. 3.4) until the time of ovulation (September/October; Fig. 3.5). Not all follicles reach preovulatory size and not all follicles of preovulatory size ovulate, but, in typical cycles, all those which do not ovulate in a given season become atretic and are reabsorbed. It is uncommon in *Pseudemydura* (although it may happen rarely in captivity: Kuchling and Bradshaw 1993) that vitellogenic follicles persist until next year's breeding season. The same pattern of vitellogenesis seems to occur in *Emydura macquarii* (Chessman 1978), *Emydura krefftii* (Georges 1983), *Chelodina longicollis* (Parmenter 1976, 1985; Chessman 1978) and *Chelodina oblonga* (Kuchling 1988b). The southernmost living species of the Pelomedusidae have not been sufficiently studied, but there may be differences in the dura-

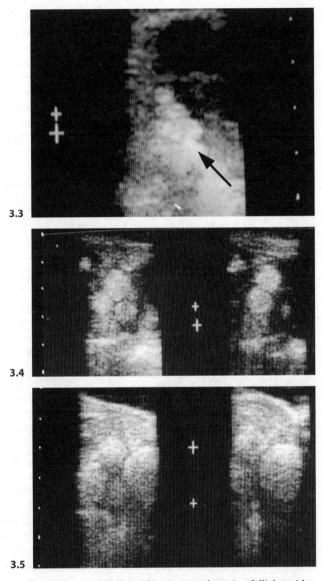

3.3

3.4

3.5

Fig. 3.3. *Pseudemydura umbrina.* Ultra-sound tomographic picture of ovarian follicles with a diameter of 4 mm (*arrow*) by February, early vitellogenesis; callipers (*white crosses*) indicate 4 mm, the scale (*white dots*) represents centimetres

Fig. 3.4. *Pseudemydura umbrina.* Ultra-sound tomographic picture of vitellogenic follicles with a diameter of 5–9 mm, April; callipers (*white crosses*) indicate 7 mm, the scale (*white dots*) represents centimetres

Fig. 3.5. *Pseudemydura umbrina.* Ultra-sound tomographic picture of preovulatory follicles with a diameter of 18 mm, September; callipers (*white crosses*) indicate 18 mm, the scale (*white dots*) represents centimetres

tion of vitellogenesis between temperate zone species of the suborders Cryptodira and Pleurodira.

Being a long-term process in chelonians, vitellogenesis frequently proceeds during periods of dormancy when energy stores, rather than energy directly assimilated from food, have to be utilised for yolk production. For example, in *Chrysemys picta* in North America, about 50% of the energy is allocated to the follicles during late summer and autumn and comes directly from harvested resources. The additional energy to complete follicle enlargement in spring comes mostly from stored body lipids (Congdon and Tinkle 1982). In the chelid turtle *Pseudemydura umbrina*, the pattern of energy allocation to follicles from stored and harvested resources takes place in the opposite way: vitellogenesis is initiated during aestivation and more than half of the energy input into the follicles occurs during summer and autumn when the animals are dormant and do not feed. Follicles continue growing during winter when the animals become active and start feeding, but, at this time, food utilisation is limited by low temperatures (King et al. 1998). In *P. umbrina*, energy allocated to follicles during summer and autumn presumably comes entirely from stored body lipids. *Pseudemydura umbrina* has to harvest most of the energy re quirements for the entire year during spring, between September and early November, when water temperatures are high enough for food processing and food is concentrating in the shrinking water bodies of the swamps (Kuchling and Bradshaw 1993).

In at least some chelonians, vitellogenesis seems to be sensitive to stress. Transfer from the wild into captivity can seriously disturb ovarian cycles in *Pseudemydura umbrina*: females taken into captivity may not initiate the following vitellogenetic cycle and females collected from the wild with large vitellogenic follicles typically do not ovulate – their follicles become atretic and are reabsorbed (Kuchling and Bradshaw 1993). Vitellogenesis is also sensitive to stress in wild *P. umbrina*: a new fence which a female cannot get through is enough to halt vitellogenic growth of follicles. Studies which combine data from wild and captive-maintained turtles should, therefore, be treated with caution. For example, Callard et al. (1978) described a slight shrinking in size of vitellogenic follicles in *Chrysemys picta* over the summer months and their resumption of growth during autumn; however, since intermittent shrinking of oocytes during vitellogenesis has not been observed in any other study of wild chelonians, it may have been an artefact due to the mixing of wild-caught turtles and individuals which were maintained in captivity for various times. Data on stress effects on vitellogenesis are still limited for chelonians, but captivity-induced stress generated massive retrogression of vitellogenesis in tropical *Anolis* lizards after as little as 4 days (Morales and Sanchez 1996).

3.2.3 Oocyte Maturation and Ovulation

In most vertebrates, including reptiles and chelonians, the oocytes are arrested in prophase I during the vitellogenic growth period and resume meiosis near or at the end of their growth. The oocytes must complete the first meiotic division to become fertilisable. The process in which prophase I-arrested oocytes resume meiosis and reach the second meiotic metaphase is called maturation. At the onset of meiosis the membrane surrounding the swollen nucleus (the germinal vesicle) is dissolved. The following process involves chromosome condensation, assembly of the first meiotic spindle and extrusion of the first polar body. In most vertebrates, meiosis is again arrested at the metaphase II stage, but this has not been investigated in chelonians. Shortly thereafter the ova are ovulated. The meiotic process leading to extrusion of the second polar body is resumed again at the time of fertilisation, immediately after sperm penetration (Nagahama 1987). According to Guraya (1989) the terms ovum maturation or meiotic maturation involve the completion of meiosis, of both successive meiotic divisions which extrude two polar bodies. Maturation and fertilisation of ova, however, is not well studied in chelonians and details of these events are still unknown.

Pseudemydura umbrina females which are transferred from the wild into captivity during late vitellogenesis typically do not ovulate, but all follicles become atretic and are reabsorbed. However, if transferred close to the time of ovulation, females may ovulate 7–14 days after being taken into captivity. In *P. umbrina*, ovulation is triggered by a physiological preparation that cannot easily be aborted 1 or 2 weeks prior to the actual event (Kuchling and Bradshaw 1993). The preovulatory changes in the follicular wall which concur with oocyte maturation may represent this "physiological preparation" which seems to determine if a follicle ovulates or becomes atretic. Oocyte maturation and ovulation are closely linked.

Ovulation is the process by which the ovum is released into the body cavity by the rupture of the follicle wall. I do not know any detailed, published descriptions of the external appearance of ovulation in any chelonian. Preovulatory follicles of several lizards, the alligator and *Chrysemys* show a circular, relatively avascular stigma which is surrounded by dilated blood vessels. A few small vessels radiate from the surrounding rosette of vessels towards the center of the stigma, like spokes on a wheel (Jones 1987). Preovulatory changes or degeneration in the follicular wall have been described for the lizard *Anolis carolinensis* (Iguanidae: Laughran et al. 1981), but not yet for any turtle. In general, the oocyte is surrounded by the vitelline membrane or zona pelludica which is formed by both the oocyte and the follicle cells, with the latter forming a peripheral layer of mucopolysaccharides. Although different definitions have been used, the zona pelludica is the primary egg membrane (Blüm 1986). The innervation of follicles and the role of enzymes in reptilian ovulation have not been studied. It is also not known if follicular contractions play a role during ovulation in reptiles, but contractile tissue has been demonstrated in the ramifying chordae of the ovary of *Trachemys scripta elegans*, with desmin-containing fibres branching throughout the medulla into the ovarian cortex where they enclose follicles and become fixed to the surface of the theca with the exception of the future rupture site, giving a particular aspect to the follicular stigma region (Callebaut and Van Nassauw 1987; Van Nassauw and Callebaut 1987). This makes the involvement of contractions of the follicular wall during ovulation likely.

Ovulation seems to be a relatively quick event and is, therefore, unlikely to be frequently observed during ultra-sound scanning. Preovulatory follicles appear to be perfectly spherical, whereas ova, once they enter the oviducts, appear elongate or elliptical with very clear outlines (see Sect. 2.2). Intermediate stages between these clearly distinct images are rarely seen. Daily ultra-sound examinations of captive *Pseudemydura umbrina* females at about the time of ovulation over several reproductive seasons (G. Kuchling and D. Burford, unpubl.) enabled us to observe, over a period of 2 days prior to the first detection of ova in the oviducts, pear-shaped ova (Fig. 3.6) and ova which seemed to have constrictions (Fig. 3.7). Images like these have to be expected if the ruptured area of the follicular wall is smaller than the diameter of the ovum, and if contractions of the follicular wall play a role during the extrusion of the ovum from the follicle. Those images may, therefore, represent different stages of ovulation, when the ovum leaves the ruptured follicle.

In most chelonians all follicles of a prospective clutch of eggs ovulate during a relatively short time span: in *Pseudemydura umbrina*, ultra-sound scanning revealed that this event takes place over 12–24h, but this time span may even be shorter. Mahmoud and Licht (1997) suspected ovulation takes 24–48h in

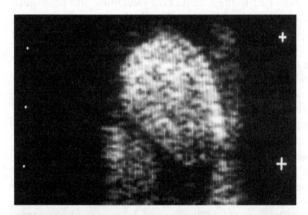

Fig. 3.6. *Pseudemydura umbrina.* Ultra-sound tomographic picture of follicle just prior to ovulation; callipers (*white crosses*) indicate 21 mm, the scale (*white dots*) represents centimetres

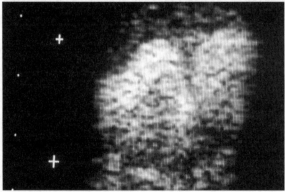

Fig. 3.7. *Pseudemydura umbrina.* Ultra-sound tomographic picture of follicle during ovulation; callipers (*white crosses*) indicate 21 mm, the scale (*white dots*) represents centimetres

Chelydra serpentina. Once the shell membrane is secreted, dissection, ultra-sound scanning or radiography of gravid chelonians typically reveal that all eggs in the oviducts are at the same stage of egg shell development. The only indication of an exception to this "simultaneous" type of ovulation in chelonians was reported by Sarkar et al. (1995) for *Lissemys punctata*, in which the gravid uterus contained both soft- and hard-shelled eggs which represented different stages of egg development within the same oviduct; they also reported that X-ray examinations demonstrated a gradual increase in the number of eggs within the uterus as time elapsed. Although radiography does not facilitate the monitoring of ovulation or the visualisation of unshelled or thinly shelled eggs in the oviducts (which ultra-sound scanning does), their findings may indicate sequential ovulation. The distinction between the "simultaneous" and the "sequential" type of ovulation in chelonians, however, is not necessarily clear cut: these types may rather represent extremes on a time continuum from a few hours to several days.

3.2.4 Luteal Phase

After ovulation the ruptured, empty follicle changes into a luteal structure (Fig. 3.8). The physiological role of the corpus luteum in reptiles was controversial for a long time, and its action may vary between groups and species, but it is now well established that all reptiles develop true corpora lutea whose life span is closely correlated with the retention of eggs or young in the female genital tract (Xavier 1987). Klicka and Mahmoud (1977) demonstrated that ovariectomy of gravid *Chrysemys picta* shortened the time to oviposition, which is similar to the results of ovariectomy in oviparous lizards (Licht 1984).

Cyrus et al. (1978) studied the corpus luteus of *Chelydra serpentina*: after ovulation, the granulosa cells of collapsed follicles proliferate and hypertrophy, fill the follicular cavity, and transform into luteal cells. Their fine structure reveals a close association of lipid droplets with abundant endoplasmatic reticulum and mitochondria. Fibroblasts of the theca interna, along with blood vessels,

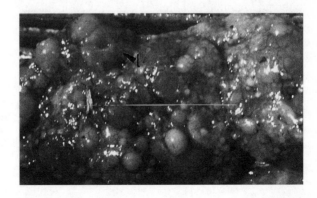

Fig. 3.8. *Erymnochelys madagascariensis.* Ovary during the breeding season with various size classes of vitellogenic follicles and corpora lutea (*l*)

invade the luteal cell mass and form septae which isolate the luteal cells into clusters. The degeneration of the corpus luteum involves massive fibrosis from the theca interna; the luteal cells disintegrate and are replaced by connective tissue. A structure forms which is comparable to the corpora albicantia of the mammalian ovary. The corpora albicantia ultimately becomes part of the ovarian stroma in which scars remain visible for some time: in *Erymnochelys madagascariensis*, for example, they remain macroscopically visible for more than half a year (Fig. 3.9).

3.2.5 Oviducal Period

The oviducal morphology of adult chelonians changes during the reproductive cycle. In *Chrysemys picta*, submucosal glands in the uterine and glandular segments were most prominent in preovulatory and postovulatory turtles and regressed after oviposition; these changes correlated with variations in the muscularis layer, the number of uterine epithelial blebs, oviducal vascularity and the presence of eosinophils in cervical segment cross sections (Abrams-Motz and Callard 1991). Some variability in the function of oviducal segments may occur, since epithelial blebs have only been observed in the uterine segment in *C. picta*, whereas in *Gopherus polyphemus* (Palmer and Guillette 1988) and in *Lissemys punctata* (Sarkar et al. 1995) they have been observed only in the infundibular region.

Ovulation sets ova free into the body cavity, but they seem to enter quickly the infundibulum section of the oviducts via the ostia abdominale. Extrauterine migration is common, and ova from one ovary may go into either oviduct (Legler 1958; Duda and Gupta 1982). This process, however, has not been directly observed in chelonians. Fertilisation presumably occurs immediately before the onset of the secretory processes in the oviduct to envelop the ovum, but the timing of fertilisation has not been studied in any chelonian. As the ova pass through the tuba uterina (or glandular segment or pars albuminifera) the clear

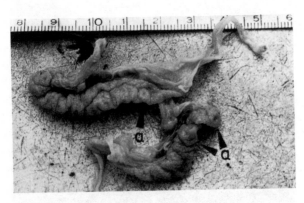

Fig. 3.9. *Erymnochelys madagascariensis.* Regressed ovary (formalin preserved) after the breeding season. *a* Corpora albicantia; no vitellogenic follicles

albuminous material is secreted around the yolk. Ultra-sound scanning of *Pseudemydura umbrina* females revealed that, once ova enter the oviducts, the albumen layer is secreted in 12–36 h (Fig. 3.10) and a shell membrane immediately thereafter (Fig. 3.11). Histological investigations of oviducts in *Gopherus polyphemus* (Palmer and Guillette 1990) and *Lissemys punctata* (Sarkar et al. 1995) support the observation that the egg albumen is secreted very rapidly and only when ova travel through the glandular segment. The rapidity of these

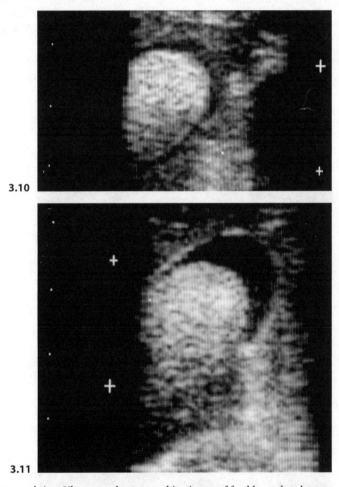

Fig. 3.10. *Pseudemydura umbrina.* Ultra-sound tomographic picture of freshly ovulated ovum in oviduct, start of albumen secretion (anechoic area surrounding echodense yolk); callipers (*white crosses*) indicate 17 mm, the scale (*white dots*) represents centimetres

Fig. 3.11. *Pseudemydura umbrina.* Ultra-sound tomographic picture of ovum in oviduct, start of shell membrane secretion, 48–60 h after ovulation; yolk echodense, albumen anechoic, thin shell membrane echodense; callipers (*white crosses*) indicate 19 mm, the scale (*white dots*) represents centimetres

45

processes is the reason why the intermittent stages between preovulatory follicles and thin-shelled oviducal eggs are rarely ever directly observed during dissections, even in studies for which several thousands of females were sacrificed (Legler 1993).

During passage of the eggs through the isthmus or intermediate segment the shell membranes are added in a series of layers around the yolk and albumen. The shell membrane thickens over a few days (Fig. 3.12) and two distinct fibrous organic layers are secreted. Then, calcium starts to be laid down in the uterine section. In *Chelonia mydas*, calcification of the shell commences approximately 72 h following ovulation (Miller 1985). The thickness and structure of the mineral layers differ between species: chelonian eggs may be parchment-shelled (pliable), hard-expansible or brittle-shelled (see Sect. 8.1). *Pseudemydura umbrina* has brittle-shelled eggs, the calcium layer of which reflects most of the ultra-sound during ultra-sound examinations and, typically, obscures the internal yolk and albumen layer (Fig. 3.13).

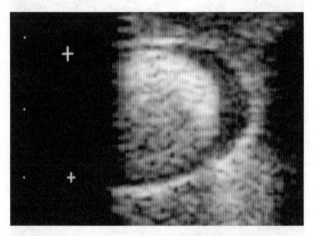

Fig. 3.12. *Pseudemydura umbrina.* Ultra-sound tomographic picture of soft-shelled egg in oviduct, shell membrane secretion completed; callipers (*white crosses*) indicate 18 mm, the scale (*white dots*) represents centimetres

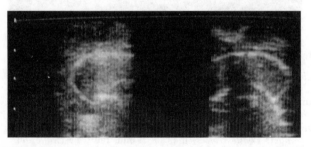

Fig. 3.13. *Pseudemydura umbrina.* Ultra-sound tomographic picture of fully calcified egg in oviduct; the scale (*white dots*) represents centimetres

Eggs move one after the other through the different sections of the oviducts and large clutches will obviously need more extended time spans for albumen and shell membrane secretion than small ones. In *P. umbrina*, which typically produces clutches of three to five eggs, I rarely observed during one ultra-sound examination ova in different stages of albumen secretion. This indicates that eggs pass through the different stages more or less simultaneously. More time delays to reach the same stage may occur in oviducal eggs of chelonians with larger clutches, but generally the oviducts contain eggs of the same developmental stages. Fifty to sixty eggs may undergo calcification at one time in each oviduct of *C. mydas* (Miller 1985). However, in *Lissemys punctata*, the sequential mode of ovulation (as discussed in Sect. 3.2.3) is paralleled by, and was actually inferred from, a gradual order of egg formation in the oviducts (Sarkar et al. 1995).

The oviducal period is variable among species, and some chelonians are capable of retaining eggs for long periods when reasonable nesting conditions are lacking. Most studies did not monitor the exact times of ovulation, but in multiclutch species nesting intervals give an indication of the oviducal periods. Moll (1979) tabled internesting periods for various chelonian species and found a range between 9 and 28 days. At the lower end of the scale is, for example, the leatherback turtle, which has a mean internesting interval of 9.6 days on the US Virgin Islands (Boulon et al. 1996) and, depending on spring tides which influence the interval, means of 9.76 or 10.01 days in French Guiana (Girondot and Fretey 1996).

In an ultra-sound study of the oviducal period of *Pseudemydura umbrina* that were kept in outdoor enclosures in their natural climate the time of ovulation was detected with an accuracy of ±24h (G. Kuchling and D. Burford, unpubl.). *Pseudemydura umbrina* typically lays only one clutch of 3–6 eggs per year and the oviducal period ranges from 35 to 49 days. Very rarely, large *P. umbrina* females produced a second, smaller clutch. In two such cases in 1995, the oviducal period was 31–32 days. The oviducal period may vary between years and between successive clutches of individual females in one season. Overall, *P. umbrina* seems to have a slightly longer oviducal retention time than many other freshwater turtles.

An X-ray study of gravid females of several North American freshwater turtles revealed maximal clutch retention times for wild *Trachemys scripta* to be 23–39 days and for wild *Kinosternon subrubrum* 26–50 days. However, prolonged egg retention in the oviducts for several months seems to be a normal pattern in the chicken turtle *Deirochelys reticularia*: wild gravid females regularly retain clutches of calcified oviducal eggs from autumn to the following spring, from 4 to 6.5 months (Buhlmann et al. 1995). Occasionally, *Chelodina steindachneri* in arid Western Australia may also retain oviducal eggs for several months (see Sect. 5.4). Wild desert box turtles *Terrapene ornata luteola* may also retain eggs for at least 50 days (Nieuwolt-Dacanay 1997). These (facultative) long egg retention times appear to allow these species to wait for good nesting conditions in unpredictable environments (see Sect. 5.4).

The two sea turtle species of the genus *Lepidochelys* have internesting periods of 14–75 days, with the average being approximately 28 days, in contrast to all other marine turtles in which internesting intervals range from 9 to 15 days. These differences in internesting periodicity are not due to interspecific differences in ovulatory cycles, but to differences in egg retention times. *Lepidochelys olivacea* and *L. kempii* are both arribada nesters, meaning that, during the nesting season, females emerge once a month synchronously to nest at beaches en masse with hundreds to thousands of conspecifics. A radio-tracking and ultra-sound scanning study at Nancite Beach, Costa Rica, revealed that, in one instance, female *L. olivacea* delayed oviposition in response to a period of very heavy rainfall. They retained oviducal eggs for 63 days and emerged synchronously to oviposite at the beach after the rain had ceased (Plotkin et al. 1997). In the genus *Lepidochelys*, egg retention may normally facilitate nesting synchrony, but also enables females to delay oviposition when environmental conditions are unsuitable.

Captive Galapagos tortoises *Geochelone nigra* may retain eggs until the next breeding season (Noegel and Moss 1989; Casares 1995), but it is not known if this also occurs in wild populations. In many chelonians egg retention for several weeks or months beyond the normal nesting time has been reported due to pathological or environmental influences in captivity (Ewert 1985). However, prolonged egg retention in captivity often causes pathologically thick eggshells, a phenomenon which has not been observed in wild *D. reticularia*, *C. steindachneri*, *T. o. luteola* or *L. olivacea*.

Oviposition terminates the oviducal phase of the eggs. The behaviours associated with nesting and oviposition are discussed in Section 4.3. Occasionally it may happen that one or two eggs fail to be deposited with the remainder of the clutch and are retained for some time in the oviducts. In *Pseudemydura umbrina*, such eggs are then simply dropped into the water a few days after nesting. I have found single, normally shelled eggs on the bottom of pools in *Chelodina steindachneri* habitat during their nesting season, suggesting similar events. I have also observed retention, after nesting, of two eggs in the oviducts of a wild *Erymnochelys madagascariensis* and similar retention of single eggs has been reported for Indian turtles of the genus *Kachuga* (Gupta 1987).

3.2.6 Regression and Follicular Atresia

At the end of a nesting season the ovaries regress. Even in year-round (aseasonal) breeding species, ovaries of individual females seem to regress at some stage: seasonality in female reptiles is rather defined at the level of the population than the individual; if seasonality is lacking, all stages of the ovarian cycle should be represented in approximately equal proportions in the population at all times of the year, but some individuals will still appear non-reproductive in each sample (Licht 1984). Although some chelonian species at low latitudes reputedly do nest year round, there is little evidence for continuous, truly acyclic ovarian activity and breeding in any chelonian species.

The degree of regression and the length of the regressive period may substantially differ among different chelonian species and even between populations of one species in different climatic zones. Ovaries are not necessarily inactive during this time. Follicular atresia commonly occurs and in many species oogenesis (oogonial mitosis) may start immediately after the breeding season (see Sect. 3.2.1). Moll (1979) calls the time span between periods of follicular enlargement, when the ovaries are at minimal size, the "latent period" and does not use the term regression.

In chelonians with multi-year ovarian cycles the regressive or latent phase of the ovaries extends over longer time spans than in annually breeding species. *Erymnochelys madagascariensis* shows a biennial ovarian cycle. After a nesting season from September to December/January, the ovaries regress and are latent until October/November when vitellogenesis starts again in preparation for the breeding season of the following year (Kuchling 1993a).

Typically *Pseudemydura umbrina* is an annual breeder and nests between late October and early December. New batches of growing vitellogenic follicles (>3 mm) can again be observed by ultra-sound scanning during January/February (see Fig. 3.3): the regressive or latent period of the ovaries only seems to be short. After the nesting season, remaining enlarged follicles become atretic and start to shrink (Fig. 3.14). Individual atretic structures may remain visible by ultra-sound for up to half a year, but others seem to disappear within several weeks.

The causes and functions of follicular atresia are still not well understood. In all vertebrate groups, oocytes at various stages of their development are eliminated from the ovaries other than by ovulation. In birds and mammals atresia of small, previtellogenic follicles is extensive and may be important in limiting the number of follicles that enter vitellogenesis and that ovulate during each cycle. Atresia of smaller follicles is not as common in the reptilian ovary. Atresia is more frequently seen in follicles which already have a polymorphic follicular epithelium, as well as in vitellogenic and fully grown follicles that have failed to ovulate. Follicular atresia in reptiles may have a function in limiting the number of eggs produced per female during a breeding season: in chelonians,

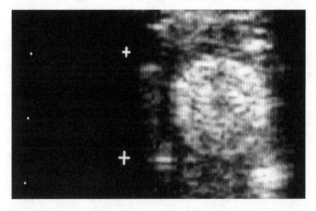

Fig. 3.14. *Pseudemydura umbrina.* Ultra-sound tomographic picture of a follicle in early atresia; variable echodensity indicates the degeneration of yolk globules; callipers (*white crosses*) indicate 16 mm, the scale (*white dots*) represents centimetres

more oocytes commonly enter vitellogenesis than are needed for the species-specific number of eggs ovulated per breeding season. Little is known about what differences exist among reptilian follicles at a similar stage of development that determine their future maturation or atresia; both extraovarian and intraovarian factors may play roles in follicular selection (Guraya 1989), but no data exist for chelonians.

The process of follicular atresia has been studied histologically in a number of lizards and snakes, but not in chelonians. During atresia of large vitellogenic follicles, the follicles shrink progressively and phagocytes which originate from follicular epithelium cells engulf and digest both the fatty and protein yolk elements. In the snake *Bungarus caeruleus* the complete resorption of large vitellogenic follicles in atresia takes several months (Guraya 1989). Ultra-sound images of chelonian vitellogenic follicles in early atresia are characterised by slight deviations from clearly spherical shape and sometimes slightly increased echodensity. Later stages of atresia reveal dense structures as well as anechoic holes and caverns, reflecting the degeneration of yolk globules and the progressive removal of yolk constituents (Fig. 3.15). During extensive ultra-sound examinations of atretic follicles in *Pseudemydura umbrina, Geochelone radiata* and *G. nigra* I found no direct indication for bursting atresia with extrusion of yolk into the stroma, as it commonly occurs in birds and as it has been described in some squamates (Guraya 1989). However, occasionally large follicles in *P. umbrina* females become invisible on the ultra-sound images within a time span of days. Yolk which diffuses into surrounding tissues may not have clearly distinguishable ultra-sound characteristics, since the method relies on relatively clear boundaries between bodies of different echodensity. The disappearance of large atretic follicles in a matter of days during ultra-sound monitoring of *P. umbrina* individuals could be related to bursting atresia.

3.3 Testicular Cycle and Sperm Production

Within the vertebrates, the basic pattern of spermatogenesis is very similar. Spermatozoa never develop alone or in isolation; the different steps of division

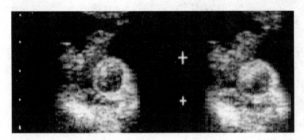

Fig. 3.15. *Chelodina steindachneri.* Ultra-sound tomographic picture of follicles in late atresia; variable anechoic holes indicate the removal of yolk constituents; callipers (*white crosses*) indicate 15 mm, the scale (*white dots*) represents centimetres

and differentiation take place in a syncytium of clonal cells which arise because the division of the cells is incomplete. Spermatogenesis only takes place in co-operation with accessory somatic cells. A difference between the amniotes and the anamniotes exists in the compartment where spermatogenesis takes place: anamniotes show cystic spermatogenesis in which cysts that are built of somatic cells constitute the primary compartment of spermatogenesis, with small or large numbers of cysts filling the tubule or ampullae (the secondary compartment); all amniotes including the chelonians show acystic spermatogenesis in association with Sertoli cells that takes place within the seminiferous tubules, which in this case represent the primary compartment (Blüm 1986).

The terminology of some components and processes of the testis–epididymis complex in reptiles differs between authors: the seminiferous tubules are also called tubuli seminiferi or tubuli contorti; these connect to the ductuli efferentes or ductuli efferentia or vasa efferentia which, in chelonians, form a rete testis between testis and epididymis (some squamates may instead have a single ductus efferens: Fox 1977); they lead into the ductuli epididymides (singular: ductulus epididymidis; some authors also differentiate between ductuli epididymides I and II) which open into the ductus epididymidis (also called ductus epididymis); the ductus deferens or vas deferens connects the epididymis to the cloaca. The terminology of the stages of spermatogenesis is: gonial or spermatogonial proliferation, spermatocytogenesis, spermiogenesis which is also called spermiohistogenesis or spermioteleosis, and spermiation.

The lining of the tubuli seminiferi is made up of the somatic Sertoli cells and of the different developmental stages of the germinal line cells. The Sertoli cells mediate between somatic influences such as hormones and the germinal line cells. The fully developed spermatozoa are released into the lumen of the tubules from where they discharge into the efferent duct system.

In tropical reptiles spermatogenesis may be continuous, but until now truly continuous spermatogenesis has not been demonstrated for any chelonian species. However, few tropical chelonians have been studied on a year-round basis. Two basic types of seasonal spermatogenetic cycles are found in reptiles: in the prenuptial pattern sperm are produced immediately before or during the mating season, whereas in the postnuptial pattern sperm are produced after the completion of the mating season and stored until the following season; a mixed pattern occurs when spermatogenesis begins in one year and proceeds until the breeding season in the next year (Saint Girons 1963). The endocrine activity of the reptilian testis is not always synchronous with the spermatogenetic activity (see Chap. 7). Chelonians generally show postnuptial spermatogenesis, which means that sperm may be stored for prolonged periods in the male urogenital system until used to inseminate a female. Only the loggerhead sea turtle (and possibly also other sea turtles) seems to show prenuptial spermatogenesis (Wibbels et al. 1990). The relationship of male and female cycles will be discussed in Section 3.5.

Definitions of the phases of spermatogenesis in reptiles differ among authors and are to some degree arbitrary. For chelonians, Moll (1979) characterised five spermatogenetic phases which generally overlap and may run parallel for pro-longed periods: germinal quiescence, gonial proliferation, spermatocytogenesis, spermiogenesis and spermiation. In an histological study, McPherson and Marion (1981b) used the abundance of the various developmental types of germ cells in the tubuli seminiferi to distinguish eight spermatogenetic stages in *Kinosternon odoratum*: (1) seminiferous tubules involuted with only spermatogonia – may have few to abundant spermatozoa in the lumens; (2) primary spermatocytes appearing – spermatogonia increasing, becoming abundant; (3) secondary spermatocytes and early spermatids abundant; (4) transforming spermatids with few spermatozoa; (5) spermatids and spermatozoa abundant; (6) spermatozoa abundant (maximum level of spermiogenesis); (7) spermatozoa abundant, but spermatids and spermatocytes greatly reduced; (8) few germinal spermatozoa – few spermatids or spermatocytes, or spermatids and spermatocytes absent, but may have abundant spermatozoa in the lumen.

In a major review, Licht (1984) proposed the following four phases as a generalised system to classify spermatogenesis in reptiles: (1) regenerative phase, the initiation of spermatogenetic recrudescence (spermatogonial divisions); (2) progressive phase, further proliferation of spermatogonia and meiotic activity up to the early spermatid stage; (3) culminative phase, active spermiogenesis with the appearance of mature spermatozoa as well as spermiation; and (4) regressive phase when no spermatogenetic activity occurs. This scheme is useful to describe and classify the high diversity in the timing of the phases in lizards and snakes with mainly prenuptial and mixed spermatogenesis. The postnuptial cycle of chelonians (and many snakes of the temperate zone) shows less variability in its seasonal timing, and the two first schemes which define more stages seem better suited to describe the various patterns.

The spermatogenetic cycle is correlated with changes in testis mass, as well as other parameters such as the diameter of the tubuli seminiferi, the height of the germinal epithelium and the overall protein concentration. Figure 3.16 summarises the annual changes of these parameters and their relationship to the developmental stages of the germ cells in *Testudo hermanni*, and indicates how they correspond to the classification schemes of Moll (1979), McPherson and Marion (1981b) and Licht (1984). The following generalised description of chelonian spermatogenesis follows the scheme of Moll (1979). It is based on various studies of temperate-zone tortoises and turtles in which the basic pattern seems to be similar, although timing and overlap of different phases may show considerable variability.

3.3.1 Germinal Quiescence

During the cool time of the year the germinal epithelium of most chelonians remains regressive and quiescent. Seminiferous tubules have a small diameter and a relatively wide lumen. The germinal epithelium is thin and contains Sertoli cells and inactive spermatogonia (Fig. 3.17). The Sertoli nuclei may migrate to some extent from the basement membrane toward the lumen. A few weeks or even months after the end of spermiogenesis there may still be some spermatozoa in the lumen, but the storage site for spermatozoa is the epididymis into which

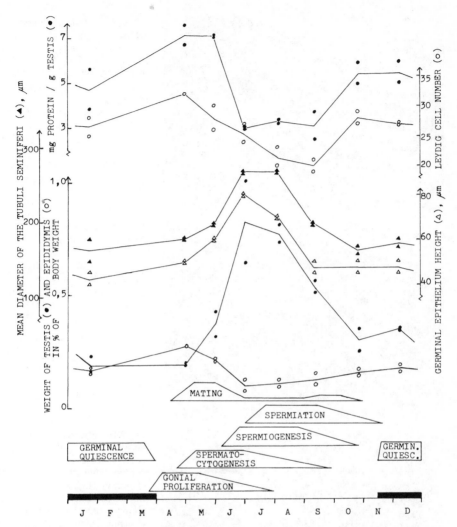

Fig. 3.16. Changes of testis and epididymis of *Testudo hermanni* in Montenegro, Yugoslavia. The scheme for the phases of the testis cycle indicates their timing and only approximately the quantitative relationship among them (Kuchling 1982a)

they are discharged. The lumen of postspermatogenetic tubules often contains cell debris from spermatocytes, spermatids, remaining spermatozoa and possibly Sertoli cells. High activity of acid phosphatase (Fig. 3.18) and β-N-acetyl-glucosaminidase (Fig. 3.19) in the regressive tubuli of *Testudo hermanni* indicates cytolytic activity. The cell masses can practically occlude the tubular lumen, owing to the reduced tubule diameter. Dubois et al. (1988) identified the presence of numerous macrophages in the lumen of regressive tubuli seminiferi of *Chrysemys picta*. In *Testudo hermanni*, lipids accumulate in the Leydig cells and close to the basement membrane in Sertoli cells (Fig. 3.20). The Sertoli cells as well

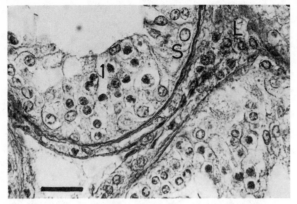

Fig. 3.17. *Testudo hermanni.* Testis, azan; germinal quiescence (January). *L* Leydig cells; *S* Sertoli cells; *1* spermatogonia; the *scale bar* represents 25 μm (Kuchling 1982a)

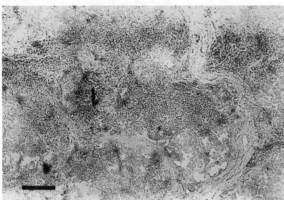

Fig. 3.18. *Testudo hermanni.* Testis, acid phosphatase; germinal quiescence (January). High activity in tubular lumen (*l*); the *scale bar* represents 50 μm

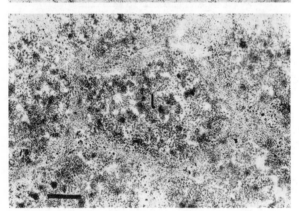

Fig. 3.19. *Testudo hermanni.* Testis, *β-N*-acetyl-glucosaminidase; germinal quiescence (January). High activity in tubular lumen (*l*); the *scale bar* represents 50 μm (Kuchling et al. 1981)

as the interstitial cells of *Mauremys caspica* show high lipid contents (Lofts and Boswell 1960). In *Pelodiscus sinensis*, the Sertoli cells of regressed tubules, as well as the Leydig cells, have a high lipid content during winter; the Leydig cells become depleted of lipoidal droplets during the mating season in spring when the germinal epithelium remains quiescent (Lofts and Tsui 1977).

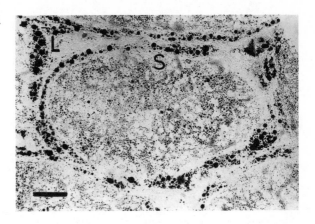

Fig. 3.20. *Testudo hermanni.* Testis, Sudan black B; germinal quiescence (January). Lipid droplets in Leydig cells (*L*) and at the base of Sertoli cells (*S*); the *scale bar* represents 50 μm

3.3.2 Gonial Proliferation

The first indication of the start of a new spermatogenetic cycle, typically during spring or early summer, is mitotic divisions of spermatogonia – the phase of gonial proliferation. The primary spermatogonia divide and produce daughter cells called secondary spermatogonia which remain joined together by cytoplasmatic bridges (Fig. 3.21). All the following steps of division and differentiation basically take place in a syncytium of clonal cells which arise because the division of the cells remains incomplete. For chelonians it is not known how many mitotic divisions secondary spermatogonia go through before they form first-order spermatocytes. In vertebrates, generally, spermatogonia go through between three and fourteen divisions (Blüm 1986). At the stage of gonial proliferation the Leydig cells of *Testudo hermanni* show a moderate to high lipid content, but, compared to germinal quiescence, the lipid content of the Sertoli cells decreases (Fig. 3.22). In *Mauremys caspica*, intratubular as well as interstitial lipids remain profuse during gonial proliferation in spring (Lofts and Boswell 1960).

Gonial proliferation may continue for many months. Since the spermatogonia are the basal mother cells which, through the process of spermatogenesis, give rise to clones of spermatozoa, the termination of gonial proliferation is the basal switch which brings spermatogenesis to an end.

3.3.3 Spermatocytogenesis

Spermatocytogenesis starts with the formation of first-order spermatocytes, the cytoplasm of which grows slightly in volume. The first-order spermatocytes then enter meiosis. Since the first maturation division is the reduction division during which the chromosomes are divided among the daughter cells so that each receives a single set, the haploid second-order spermatocytes have distinctly smaller nuclei. After the second maturation division (segregation division)

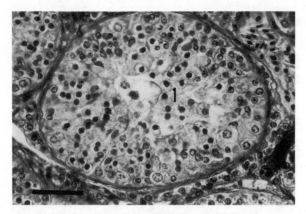

Fig. 3.21. *Testudo hermanni.* Testis, azan; gonial proliferation (March). spermatogonia (*1*) multiply; the *scale bar* represents 50 μm

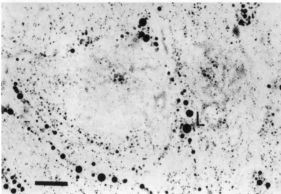

Fig. 3.22. *Testudo hermanni.* Testis, Sudan black B; gonial proliferation and early spermatocytogenesis (April). Lipid droplets mainly in Leydig cells (*L*); the *scale bar* represents 50 μm (Kuchling et al. 1981)

these become spermatids, the nuclei of which are again smaller (Fig. 3.23). The number of secondary spermatocytes is low compared to the other cell types: they are only a transitory stage. The diameter of the tubuli seminiferi, the height of the germinal epithelium and the testis mass all increase as spermatocytogenesis progresses (compare with Fig. 3.16). In *Mauremys caspica*, the cytoplasmatic lipids of Sertoli cells and Leydig cells become progressively denuded of lipids (Lofts and Boswell 1960), a process less pronounced in *Testudo hermanni* (Kuchling 1979).

3.3.4 Spermiogenesis

The final stage of spermatogenesis is the transformation of the spermatids into motile spermatozoa which involves remarkable changes in their morphology, a process known as spermiogenesis or spermiohistogenesis. Nuclear and cytoplasmatic volume changes occur and most of the cytoplasm is eventually eliminated. In vertebrates generally, spermatids are spherical in the early stages

Fig. 3.23. *Testudo hermanni.* Testis, azan; spermatocytogenesis (late May). *1* Spermatogonia; *2* first-order spermatocytes; *3* second-order spermatocytes; *4* early spermatids; the *scale bar* represents 25 μm (Kuchling 1982a)

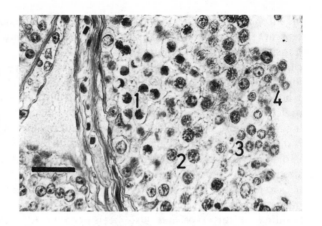

Fig. 3.24. *Testudo hermanni.* Testis, azan; spermiogenesis (early August). The round nuclei of early spermatids (*4*) condense and elongate (*5*); the *scale bar* represents 50 μm

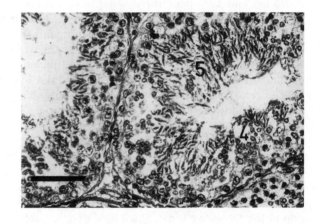

and contain a nucleus, mitochondria and the two centrioles. A Golgi apparatus then develops in the vicinity of the nucleus and forms an acrosomal vesicle enclosing an acrosomal granule which then comes to lie against the nucleus like a cap. At the same time the centrioles migrate to the periphery of the cell roughly opposite the acrosomal vesicle where one of them gives rise to a small flagellum. This centriole subsequently moves towards the nucleus, thereby prolonging the flagellum. Mitochondria aggregate at the flagellate pole of the cell and, eventually, group around the flagellum. At the same time the nucleus condenses and becomes elongate, the cytoplasm shifts and most of it is eliminated (Fig. 3.24). In the last phase the acrosome takes on its final form and the head of the spermatozoon is formed (Blüm 1986).

Sprando and Russell (1988) described the cytoplasmatic events during spermiogenesis in *Trachemys scripta*: the first step is that nuclei which are in the centre of newly formed spermatids become eccentrically located within the cell. The nuclear pole of the spermatid is situated within deep crypts of a Sertoli cell. The cytoplasm of the elongating spermatid is displaced along the non-acrosomal region of the nucleus and the proximal flagellum. Sheet-like Sertoli cell processes indent the cytoplasm of the spermatid adjacent to the nucleus and segregate and separate small packets of cytoplasm from the spermatid. During the progressive translocation of the spermatid from the deep crypts of the Sertoli cell to a position closer to the lumen, a portion of the spermatid cytoplasm is also displaced forward over the acrosomal region of the spermatid, at which stage it resembles a hood. As the spermatid is transported to the seminiferous tubular lumen, a preferential flow of cytoplasm into one area of the hood forms a forward-project-ing lobe. Before the release of the sperm into the lumen, this cytoplasmatic lobe is disengaged from the spermatid head and forms a large residual body ($2-4\,\mu m$ in diameter) which is internalised and degraded within the Sertoli cell. Medium-sized ($1-2\,\mu m$) lipid-rich cytoplasmatic lobes are also pinched from the head and neck region and small-sized ($0.5-1\,\mu m$) mitochondrial-rich cytoplasmatic frag-ments bud from the caudal head and midpiece of the spermatid. Mitochondrial budding of this type has also been seen in *Chrysemys picta*, *Kinosternon odoratum* and *Terrapene carolina* (Sprando and Russell 1988). The discarded residual bodies may float free at the rim of the tubular lumen, and are engulfed and phagocytised by the Sertoli cell. As in mammals, residual bodies form late during spermiogenesis near the time of sperm release. Upon release of the spermatozoon from the Sertoli cell into the lumen of the tubuli seminiferi, a perinuclear band of cytoplasm containing lipid and clear membrane-bound vesicles remains associated with the spermatozoon's head in *Trachemys scripta* (Sprando and Russell 1988).

Various enzymes play a role during spermiogenesis in mammals (Vanha-Perttula 1978), but have been little studied in chelonians. During spermiogenesis, the germinal epithelium of *Testudo hermanni* shows a high activity of β-N-acetyl-glucosaminidase (Fig. 3.25), an enzyme which is involved in the transformation of glycoproteins in plasma membranes. Acid phosphatase shows little activity in the germinal epithelium during spermiogenesis (Fig. 3.26), in contrast to the situa-tion in the regressive testis. Small lipid droplets are associated with spermatids and free spermatozoa in the lumen, which corroborates the findings of high cytoplasmatic lipid contents of spermatozoa discussed above; the lipid content of the Leydig cells is slightly reduced (Fig. 3.27). Surprisingly, with similar his-tochemical techniques the germinal epithelium as well as the free spermatozoa in the tubule lumen were reported to be largely lipid-free during spermiogenesis in *Mauremys caspica* (Lofts and Boswell 1960) and in the soft-shelled turtle *Pelodiscus sinensis* (Lofts and Tsui 1977), but, unfortunately, without these par-ticular results being presented in figures or plates. More data on these turtles and other chelonian families would be needed to evaluate if it is common for all chelonians to have sperm with high cytoplasmatic lipid content, or if exceptions exist.

3.3.5 Spermiation

Spermiation is the process of dissociation of fully differentiated spermatozoa from the Sertoli cells and their accumulation in the lumen of the ductuli

Fig. 3.25. *Testudo hermanni.* Testis, β-*N*-acetyl-glucosaminidase; spermiogenesis (early September). High activity in germinal epithelium (*e*) and in mass of free spermatozoa in lumen (*l*); the *scale bar* represents 50 µm (Kuchling et al. 1981)

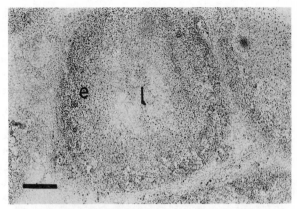

Fig. 3.26. *Testudo hermanni.* Testis, acid phosphatase; spermiogenesis (early September). Moderate activity, restricted to interstitium and basal areas of the germinal epithelium; the *scale bar* represents 25 µm

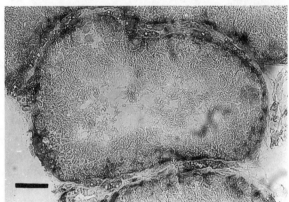

Fig. 3.27. *Testudo hermanni.* Testis, Sudan black B; spermiogenesis and spermiation (early September). Some larger lipid droplets in Leydig cells (*L*), small lipid droplets associated with spermatids in germinal epithelium (5) and free spermatozoa in lumen (*l*); the *scale bar* represents 50 µm

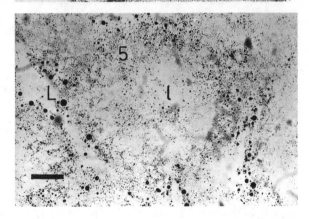

seminiferi. In *Trachemys scripta*, the total cell volume of the spermatozoon is reduced by 79% when compared with the young, round spermatid; this overall decrease is the result of an 84% cytoplasmatic reduction and a 78% nuclear reduction. The 79% total cell volume reduction falls into the range observed in various mammals, but is less than the reduction observed in Osteichthyes

(Bluegill: 87%), the bullfrog (87%) and the domestic fowl (97%: Sprando and Russell 1988).

In *Chrysemys picta* (and, presumably, other turtles) the perinuclear cytoplasm gives rise to a cytoplasmatic droplet which moves posteriorly along the midpiece following detachment of sperm from the Sertoli cell (Gist et al. 1992). This lipid-rich cytoplasm may be the counterpart of the cytoplasmatic droplet of the mammalian spermatozoon. In most mammals, however, the cytoplasmatic droplet does not contain lipid and is located at the juncture of the head and flagellum. In *Chrysemys picta* the cytoplasmatic droplet does not move down the flagellum, but remains at the midpiece (Gist et al. 1992). During spermiation, spermatozoa and their attached droplets are delivered to the epididymis. The high lipid content of the chelonian spermatozoon seems to be unique among vertebrates, and may serve as energy store for the often protracted storage time of sperm in the epididymis.

3.3.6 Maturation and Storage of Sperm in the Epididymis

In mammals it is well established that the spermatozoa undergo a maturation process during their passage through the epididymis. Among other events, the antigenic structure of the spermatozoan surface is changed during its transit. The antigenic structure of the plasma membrane depends on glycoproteins and glycolipids, which can be transformed by the sequential effect of glycosidases, including β-N-acetyl-glucosaminidase. Variable high activities of β-N-acetyl-glucosaminidase have been found in the epithelial cells in defined but, according to the species, different segments of the epididymis (Bamberg 1975).

Considerable secretory activity occurs in the epithelium of the ductuli epididymides and of the ductus epididymidis of the tortoise *Testudo hermanni* (Kuchling et al. 1981). At the time when spermatozoa are released from the testis, the ductuli epididymides show segments with varying degrees of periodic acid Schiff (PAS) reactions in the epithelium (Fig. 3.28), with varying degrees of activity of acid phosphatase (Fig. 3.29) and β-N-acetyl-glucosaminidase (Fig. 3.30) and with variable amounts of lipid droplets in the epithelium (Fig. 3.31). The β-N-acetyl-glucosaminidase and the lipid droplets are released during the passage of spermatozoa through those particular segments (Figs. 3.32, 3.33). Biochemically, the overall activity of β-N-acetyl-glucosaminidase in the epididymis also increases with the onset of spermiation in summer and remains elevated during autumn (Fig. 3.34); however, in contrast to mammals, it remains lower than the activity in the testis. The epididymis of the lizard *Lacerta vivipara*, too, shows a dramatic increase of its secretory activity as spermatozoa enter the duct and undergo maturation; secretory products include proteins which are glycosylated with N-acetylglucosamine (Ravet et al. 1991) and which bind to the heads of spermatozoa (Courty et al. 1987), suggesting maturation processes which may be comparable to those in *T. hermanni* (and mammals).

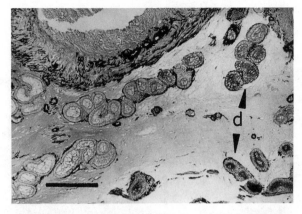

Fig. 3.28. *Testudo hermanni.* Epididymis, PAS; spermiation (October). Strong PAS reaction in epithelium and lumen of some ductuli epididymides (*d*); the *scale bar* represents 200 μm

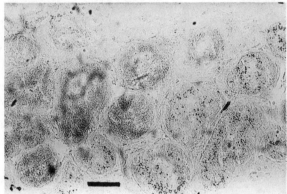

Fig. 3.29. *Testudo hermanni.* Epididymis, acid phosphatase; spermiogenesis (July). Variable strong reaction in epithelium and lumen of some ductuli epididymides; the *scale bar* represents 50 μm

Postnuptial spermatogenesis, the most common pattern found in chelonians, requires that viable sperm are maintained for protracted periods in the efferent duct system. The main sites of sperm storage in chelonians are the ductus epididymidis and, to a lesser extent, the vas deferens. The ductus epididymidis of *T. hermanni* acts over its full length as a storage site for spermatozoa and shows enzymatically no distinct differentiated segments. Corresponding to the different functional arrangement of the epididymis, the enzymatic processes which occur in distinct segments of the ductus epididymidis in mammals are, in *T. hermanni*, located in the area of the ductuli epididymides. Because of the high activity of β-N-acetyl-glucosaminidase in the testis it might also be that some of the enzymatic maturation processes have already taken place in the seminiferous tubules (Kuchling et al. 1981).

The epithelium of the ductus epididymidis of *T. hermanni* is practically lipid-free, but the sperm mass inside shows a homogenous density of lipid droplets (Fig. 3.35). This corresponds to the high lipid content of the cytoplasmatic droplets of spermatozoa in *Trachemys scripta* (Sprando and Russell 1988) and *Chrysemys picta* (Gist et al. 1992). The epithelium of the ductus epididymidis of *T.*

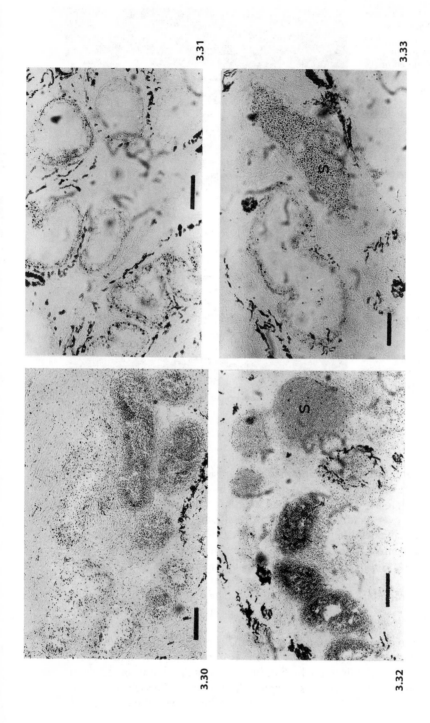

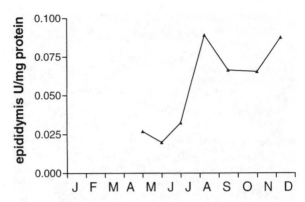

Fig. 3.34. *Testudo hermanni.* Activity of β-N-acetyl-glucosaminidase in the epididymis. *U* denotes the activity of the β-N-acetyl-glucosaminidase according to the method described by Levvy and Conchie (1966). (Kuchling et al. 1981)

Fig. 3.35. *Testudo hermanni.* Epididymis, Sudan black B, June. Epithelium of ductus epididymidis (*e*) lipid free, but small lipid droplets in mass of spermatozoa in lumen (*s*); the *scale bar* represents 50 μm

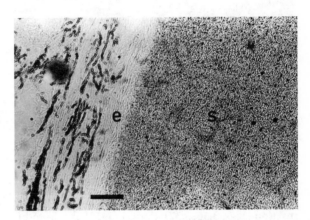

◄ ──

Fig. 3.30. *Testudo hermanni.* Epididymis, β-N-acetyl-glucosaminidase; spermiogenesis (July). Variable strong reaction in epithelium and lumen of some ductuli epididymides; the *scale bar* represents 50 μm (Kuchling et al. 1981)

Fig. 3.31. *Testudo hermanni.* Epididymis, Sudan black B; spermiogenesis (July). Variable amounts of lipid droplets in the epithelium of some ductuli epididymides; the *scale bar* represents 50 μm (Kuchling et al. 1981)

Fig. 3.32. *Testudo hermanni.* Epididymis, β-N-acetyl-glucosaminidase; spermiation (September). Strong reaction in epithelium of empty ductuli epididymides, but epithelium devoid of activity in ductuli filled with masses of spermatozoa (*s*) which show moderate activity; the *scale bar* represents 50 μm (Kuchling et al. 1981)

Fig. 3.33. *Testudo hermanni.* Epididymis, Sudan black B; spermiation (September). Lipid droplets in epithelium of empty ductuli epididymides, but epithelium devoid of lipids in ductuli filled with masses of spermatozoa (*s*) which contain many small lipid droplets; the *scale bar* represents 50 μm (Kuchling et al. 1981)

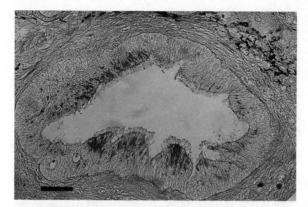

Fig. 3.36. *Testudo hermanni.* Epididymis, PAS, early August. Epithelium of empty ductus epididymidis shows areas with strong PAS-positive reaction; the *scale bar* represents 50 µm

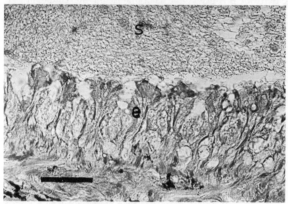

Fig. 3.37. *Testudo hermanni.* Epididymis, PAS, early December. Epithelium of ductus epididymidis (*e*), which is filled with masses of spermatozoa (*s*), shows apically a PAS-positive reaction; the *scale bar* represents 50 µm

hermanni shows some secretory activity, as indicated by a slightly positive PAS reaction (Figs. 3.36, 3.37). Furieri (1959) also found PAS-positive secretory activity in the epithelium of the ductuli epididymides and the ductus epididymidis of *Emys orbicularis*. In *Lissemys punctata*, the sialic acid and glucose content increased in the epididymis during and after spermiation (De and Maiti 1989). Changes to spermatozoa in the ductus epididymidis were described for *Chrysemys picta*, in which the cytoplasmatic droplets become detached from the sperm midpiece in a coordinated manner shortly before the commencement of autumn mating. This process is not a function of position within the ductus epididymidis, since sperm from all regions is in the same state; rather it seems to be a synchronised event and may be triggered by hormonal or other mechanisms (Gist et al. 1992). Furieri (1959) studied fresh sperm from the ductus epididymidis of *Emys orbicularis* by phase-contrast microscopy and found granules intermixed with spermatozoa which he interpreted as secretory products of the epididymis; it is, however, possible that those granules represented, at least in part, detached cytoplasmatic droplets. The loss of the cytoplasmatic droplet from the spermatozoon in the ductus epididymidis represents an additional reduction of the cell volume from that during spermiogenesis. The extratesticular plasma reduction which occurs in the ductus epididymidis a considerable amount of time after the release of sperm from the Sertoli cells may be the final maturation event for chelonian spermatozoa.

3.4 Accessory Secretions and Insemination

The total period of sperm storage in the male genital tract varies among species and between populations, and depends on the timing of spermiogenesis – which, usually, takes place during the hottest season – and the time of mating and insemination of females, which is often during the cooler seasons, in autumn, winter or spring (for details see Sect. 4.2.1). The chelonian penis was described in Section 1.3.1 and courtship behaviour will be treated in Section 4.2.3. Very little is known of the actual transfer of sperm from the male to the female, and if accessory secretions of the male are involved in the constitution of the ejaculate.

Chelonians (and crocodiles) lack the sex segment of the kidney of male lepidosaurs, the secretions of which contribute to the semen and may activate sperm (Fox 1977). However, even without evidence of seasonal or androgen-dependent nephron tubule hypertrophy, kidney products do activate or inhibit sperm motility in *Trachemys scripta*, depending upon the season: during the April breeding period, kidney homogenate cytosols increase motility of sperm removed from the ductus deferens, but inhibit motility when tested in other seasons (Garstka and Gross 1990). These results suggest that chelonians may have a functional, if not anatomical, renal sex segment, which may be the primitive condition for reptiles, with the renal sex segment of the lepidosaurs being a derived condition. The most derived state of this trait is the mammalian condition with separate glands, each with a specialised function.

The epithelial layer of the ductus deferens of *Testudo hermanni* contains cup-shaped glands with columnar cells and basal nuclei which are distinct from the general ciliated epithelial cells (Fig. 3.38). Some of the glands form crypt-like tubes with well circumscribed openings into the lumen of the ductus deferens (Fig. 3.39). Although the function of these glands has not been studied, it is very likely that their secretion contributes to the semen or the ejaculate. The whole area of accessory secretions during semen transfer from male to female chelonians is not well studied or understood.

3.5 Sperm Storage in the Female Genital Tract

The annual cycles of most organisms are synchronised so that male and female gametes mature and are released at the same time. The release of spermatozoa and ova from chelonian gonads, however, is typically asynchronous and occurs at different times of the year (the various patterns will be discussed in Chapter 5).

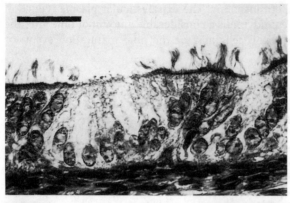

Fig. 3.38. *Testudo hermanni.* Epithelium of vas deferens, HE, May. Cup-shaped glands; the *scale bar* represents 25 μm

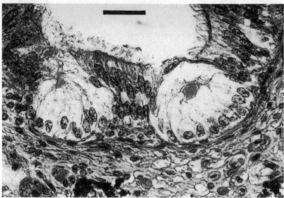

Fig. 3.39. *Testudo hermanni.* Epithelium of vas deferens, Haematoxylin-Eosin (HE), May. Crypt-like glands; *scale bar* =25 μm

Since fertilisation of the ovum, presumably, has to occur in the narrow time span between ovulation and the secretion of the albuminous envelop around the yolk – a few hours or perhaps a day (see Sect. 3.2.5) – sperm has to be stored to be available for fertilisation. The importance of the epididymis and the vas deferens for extratesticular sperm storage in males has been elaborated in the previous section. Although many chelonians mate at about the time of ovulation, this is not always the case. For example, several spring breeding species may show a mating period in autumn or winter (see Sect. 4.2.1). Cloacal lavages of freshly captured females of the spring nesting musk turtle *Kinosternon odoratum*, for instance, revealed motile sperm in 55% of autumn-caught females but in only 10% of spring-caught females (Mendonça 1987a), indicating that insemination of females in this species took place mainly in autumn. To facilitate fertilisation in those cases, sperm must be stored and kept viable for a prolonged period in the female genital tract. Some turtles can continue to lay fertilised, viable eggs for several years following isolation from males (e.g. *Malaclemys terrapin*: Barney 1922; *Terrapene carolina*: Ewing 1933; *Pseudemydura umbrina*: Kuchling et al. 1992).

Hattan and Gist (1975) reported oviducal tubules containing sperm for the turtle *Terrapene carolina* and Gist and Jones (1989) studied and confirmed oviducal sperm storage tubules between the infundibulum and the uterus in 11 other turtle species of 6 different families (Emydidae, Kinosternidae, Chelydridae, Trionychidae, Cheloniidae and Testudinidae). For some of these species these studies reconfirmed older, circumstantial evidence for sperm storage by females. Sperm storage tubules are found in a small region of the posterior portion of the egg albumin-secreting section of the oviduct. The tubules are lined by a single layer of prismatic cells which seem to be identical to the albumen-secreting cells. Ducts, restricted to the posterior albumin region, connect the tubules to the oviduct lumen and allow entrance of sperm to the tubules. The ducts are lined by ciliated and unciliated epithelium similar to that lining the entire albumen-secreting region. It seems that neither the ducts nor the secretory tubules are particularly modified for sperm storage. Sperm are stored loosely or in aggregates within the tubules and usually have their heads oriented towards the terminus, but not in contact with the epithelium. Gist and Jones (1989) identified sperm in tubules of female turtles which were isolated from males for up to 423 days.

Sperm storage structures in the oviducts of lizards, snakes (Saint Girons 1975) and birds (Bakst 1987) are located either in the infundibulum or in the lower portion of the oviduct at the uterovaginal junction. The location of the chelonian sperm storage tubules, between the infundibulum and the uterus, is unique among vertebrates. It seems that neither the ducts nor the secretory tubules are particularly modified for sperm storage.

Chapter 4

Reproductive Behaviour

According to both ethological and physiological criteria, the reproductive behaviour of the majority of vertebrates has two phases, a sexual phase and a phase of parental care (Blüm 1986). The objective of the sexual phase in chelonians is copulation and the insemination of the female. Parental care is mostly limited to nesting, the concealment of the eggs in a favourable place, which is solely the task of the female. Only a few chelonian species show nest guarding behaviour.

The distinction between those behaviours that are related to reproduction and those that are not is blurred, and to some degree arbitrary. For example, thermoregulatory behaviour may be important for the physiological condition and, therefore, for the reproductive performance of an individual turtle (King et al. 1998), but this is an indirect rather than a direct relationship and I do not include it in this account. Migration, on the other hand, may also occur in relation to food sources or the thermal environment, but it is often directly associated with breeding. I, therefore, discuss migration in this chapter.

Although the emergence of hatchlings from nests and their movement to the habitat where they grow up is not part of reproductive behaviour *sensu strictu*, I discuss it at the end of this chapter since it represents the beginning and the end of nesting movements and migrations. In migrating species, the migration basically starts when hatchlings emerge from the nest. In non-migrating species, too, hatchlings have to negotiate the same distance from the nest to the foraging habitat as the females. Since this important phase in the life of chelonians directly depends on the nesting behaviour of the females, I include it in this chapter.

A comprehensive review of behavioural patterns of chelonians is provided by Carpenter and Ferguson (1977), a review of social behaviour by Harless (1979) and of behaviour associated with nesting by Ehrenfeld (1979). The following account is not an exhaustive review of the literature on detailed descriptions of variability in stereotypical behaviour; I rather focus on certain behavioural aspects which provide insight into adaptations to different life styles and ecological requirements.

4.1 Migration

Migratory behaviour is common in sea turtles, but also occurs in a variety of other groups. The leatherback turtle *Dermochelys coriacea* is a strong, though possibly vagrant, migrator. It has pelagic feeding grounds in temperate seas and is frequently sighted in cold, northern or southern waters, but nests on tropical or subtropical beaches (Meylan 1995). Individual leatherback turtles may nest on different beaches in consecutive seasons and may even make substantial intraseasonal shifts in nesting sites (Boulon et al. 1996). Migratory behaviour of the other sea turtle species depends on the habitat, evolutionary history and life style of particular populations, but is, generally, less flexible. An example of an impressive breeding migration of the green turtle (*Chelonia mydas*) is the breeding population at Ascension Island in the south Atlantic Ocean which has its feeding ground in Brazilian waters; the turtles must cross 2200 km of open ocean to reach Ascension Island to breed. Carr and Coleman (1974) hypothesised that the Brazil–Ascension pattern evolved when the distance to be travelled was slight, and that the present situation resulted from gradual sea floor spreading due to continental drift during the early Tertiary. The Ascension turtles share the Brazilian foraging grounds with green turtles which nest at Surinam beaches. The two populations feed side by side, but migrate to separate locations to nest (Bowen et al. 1989).

Two hypothesis were originally proposed as to how sea turtles may establish their breeding migration routes. Hendrickson (1958) proposed a social facilitation model which assumes that first-time nesters follow experienced breeders to a nesting site, which they learn by olfactory and other navigation systems. The imprinting or natal homing hypothesis (Carr 1967) suggests that marine turtles may nest on their natal beaches and that the turtles learn characteristic components of their natal beach early in life. Recent molecular genetic studies of several marine turtles (*Chelonia mydas*, *Caretta caretta*, *Eretmochelys imbricata*; reviewed by Bowen 1995) support the natal homing hypothesis. Nesting populations of these species are distinct demographic units. According to this concept there is little potential for depleted nesting populations to be replenished by recruitment from other nesting locations. The beach environment, however, is not stable over thousands of years and occasional lapses in natal homing must occur to allow the colonisation of new nesting habitat. This hypothesised process may also limit the development of intraspecific evolutionary separations within each ocean basin (Bowen 1995).

The migrations of sea turtles begin at the hatchling stage and continue throughout life. Relatively little – considering the amount of research dedicated to sea turtles – is known of navigational mechanisms. For a while hypotheses focused on olfactory or chemical cues (Owens et al. 1982). Recent satellite tracking experiments have revealed that migrating adult green turtles often follow essentially straight paths to target sites from hundreds of kilometres away, even when moving across the open ocean or perpendicularly to water currents. The migration paths indicate neither zigzagging progress nor random directional fluctuation. This suggests that turtles do not rely on chemical cues during long-distance navigation (Lohmann et al. 1997; Sakamoto et al. 1997).

It is now well established that hatchlings, after they enter the sea, orient in the near-shore zone into oncoming waves (Wyneken et al. 1990). Since waves and swells move almost directly towards sand beaches where turtles nest, swimming into waves results in movement away from land and towards the open sea. In deeper water farther from land, waves no longer provide a reliable indicator of offshore direction. The hatchlings, however, have the ability to maintain seaward courses long after they swim beyond sight of land. Sun compass orientation was demonstrated in green turtle hatchlings (Fischer 1964), but it now seems that geomagnetic cues are mainly responsible for guiding the course of hatchlings travelling offshore. During their initial swimming through the wave refraction zone, hatchlings acquire a magnetic directional preference (Lohmann and Lohmann 1994), so that they continue in the same direction after entering the open sea. They are guided by an inclination compass, similar to the magnetic compass of birds, which functions by distinguishing the angle of the earth's magnetic field lines, which vary with latitude. Sea turtles are thus able to determine their latitudinal position magnetically. Their magnetic sense has the potential to indicate both direction and position and may be the key guidance system while hatchlings remain in the pelagic current systems (Witherington 1995). The pelagic stage of sea turtles is known as the "lost year", but recent studies suggest that it is more likely a "lost decade" and that it may involve considerable migrations, like movements in the North Atlantic Gyre system (Carr 1987) and transoceanic movements in the North Pacific (Bolten and Balazs 1995). After their pelagic stage, cheloniid sea turtles settle as juveniles in coastal, benthic foraging habitats. It takes cheloniid turtles 40–60 years to attain sexual maturity (Bjorndal and Zug 1995), but females of the larger leatherback turtle mature in 13–14 years (Zug and Parham 1996).

Despite the recent progress in resolving some guidance mechanisms in hatchlings there is still a lack of tenable theories on the imprinting mechanisms and how maturing turtles return, after decades, to their natal beach, which may be a little island in the open sea. Lohmann et al. (1997) hypothesised that adult sea turtles may integrate magnetic field inclination and intensity with bicoordinate magnetic maps that allow journeying between nesting beaches and foraging sites. Owens (1997) speculated that olfactory imprinting occurs just prior to hatching and at 4 or 5 days of age as the turtles enter the water, since studies done in his laboratory showed that corticosterone levels are low during these phases and birds appear to be able to imprint optimally when their corticosterone levels are low. The stage when the turtles enter the water with low corticosterone levels would obviously also allow imprinting style learning of a magnetic map, an hypothesis which Owens (1997) did not consider. Sakamoto et al. (1997) revived the concept of sun compass orientation and speculated that underwater solar information is an important factor, since the sunrise and sundown hour is correlated with long-range migration in the open ocean, but total geomagnetic field values are not correlated with the path. All three concepts have not yet been proven, but may occur concurrently. At present, the open-ocean navigational abilities of sea turtles still remain a mystery. Carr (1995) described this fact as "an embarrassment to science".

To some degree it is a question of judgement whether a population does or does not migrate. Green turtles which congregate at particular sites to mate and nest and which spend the time between their reproductive bouts in distant feeding grounds are clearly migratory. In green turtles, requirements for nesting are generally different from feeding site requirements. Being largely herbivorous, green turtles rarely find adequate plant food along the exposed coasts on which good nesting sand piles up. However, where palatable algae tolerate open-shore conditions, as in the Galapagos Islands and parts of the Hawaiian Archipelago, green turtle colonies are classified as nonmigratory (Carr 1995). Since the pelagic juvenile stage is still maintained in these populations, the main criterion for this sea turtle migration concept seems to be the distance between the breeding and foraging sites. It has been suggested that for some species, such as the hawksbill turtle *Eretmochelys imbricata*, nesting and feeding requirements do not impose the necessity of migration (Carr 1952), but recent data suggest that some hawksbill populations may still migrate extensively. Indications for at least some migratory behaviour have now been found in all species of sea turtles (Meylan 1995).

Compared with sea turtles, migrations of other chelonians are rather modest. Several large river turtle species aggregate in areas with sand banks for mating and nesting, a pattern directly comparable to the breeding migrations of sea turtles. Two well known examples are the tutong *Batagur baska*, a large Southeast Asian river turtle which lives in brackish estuaries of large rivers and migrates up to 80 km upriver to nest colonially on sand banks; and the arrau turtle *Podocnemis expansa* (the largest pleurodire turtle, widely distributed in the Amazon and Orinoco river systems), which congregates for nesting in certain areas with extensive sand banks (Pritchard 1979a; Ernst and Barbour 1989).

In all the cases of migration presented above, the turtles leave the area where they hatch and, once they reach sexual maturity, only return to breed. No single definition for migration is, however, universally accepted. For some other chelonian species or populations it is a question of definition if they are classified as migratory or not. For fish, migration in rivers can, for example, be defined as the periodic movement of a significant fraction of the population which results in an alteration of their habitats (Northcote 1984). An even wider definition of migration is "the act of moving from one spatial unit to another" (Baker 1978). If the distance between the spatial units is not critical, the number of chelonian species which can be classified as migratory increases. Baker (1992) even went so far as to suggest that the division of animals into migrants and nonmigrants and the attempts to define migration are scientific misconceptions, since all animals are born in a place determined for them by their mother and die at a place determined by their own migrations – this path through time and space being their "lifetime track". This expansion of the term migration would mean that any chelonian species or individual is, a priori, migratory. This approach, however, neither clarifies nor helps to understand the phenomenon in question. Berthold (1992) restricted the term migration to events where creatures move at least beyond the boundaries of their normally inhabited home ranges (as opposed to "spatial unit" which is much more open to arbitrary definition).

This short – and by no mean comprehensive – discussion of some concepts of migration demonstrates that there is no universal agreement on how or even if migration should be defined. Practical distinctions of different migratory events include calculated versus non-calculated, facultative versus obligatory, return versus removal, and exploratory migrations (Baker 1978). Dingle (1992) distinguished between one-way migration, nomadic migration, return (to and fro) migration and migration circuits. According to this concept, most sea turtles fall under migration circuits, with the separation of life cycle stages (egg incubation, pelagic growth phase and benthic subadult and adult feeding) among areas joined in a migration circuit; the pelagic leatherback turtle may, in addition, show tendencies of nomadic migration. The flatback sea turtle *Natator depressus*, which does not have a pelagic phase (Walker and Parmenter 1990), the tutong and the arrau turtle exhibit return (to and fro) migration. One-way migration would be the colonisation of new habitats. In respect of colonisation, it is interesting to note that male chelonians show much stronger tendencies for long-range movements and invasion of new habitats than either females or juveniles (Tuberville et al. 1996).

For oviposition, females of most freshwater turtles leave their freshwater habitat and move to a terrestrial nesting site – which, even if adjacent, is a different habitat. I would not generally describe nesting movements as migration, but some authors do and there is a wide range of gradation in how far females move to nest, or if they congregate, and the line between when migration occurs or does not is arbitrary and relative. For example, nonmigratory green turtle populations (Carr 1995) may show more substantial spatial and temporal movements than the rather conventional nesting movements of *Chrysemys picta* which were described as migration (Congdon and Gatten 1989). Remarkable cases of extensive nesting movements (over several kilometres) include the slider turtle (*Trachemys scripta venusta*) population from Tortuguero in Caribbean Costa Rica: females, which attain an extraordinary size, leave their lotic habitat, cross a peninsula, enter the sea, drift-swim southwards with the longshore current and, during the night, haul up on the sea beach to nest (Moll 1994); and the snapping turtle *Chelydra serpentina* in the northern part of its distribution: females move annually several kilometres between their home ranges and their preferred nesting sites (Obbard and Brooks 1980).

However, one important aspect, apart from the distance travelled, separates the reproductive migrations of sea turtles and of river turtles like the arrau turtle *Podocnemis expansa* from the nesting movements of many other chelonians which are also sometimes called migration. This is the fact that the former mate and copulate in the water adjacent to the nesting beaches before nesting commences, which means that males as well as females migrate away from their feeding grounds to reproduce, whereas in the other species only gravid females are on the move in search of nesting sites. I propose to use this aspect to distinguish reproductive migrations of chelonians from nesting movements.

Chelonian populations may also show seasonal shifts between habitats which are not exclusively related to breeding or nesting, some of which have also been described as migrations. Iverson (1991) described the migration of yellow mud turtles *Kinosternon flavescens* between an aquatic and an adjacent terrestrial sand hill habitat where the turtles nest, aestivate and hibernate, and emphasised the fidelity of individuals to their migration paths. Quinn and Tate (1991) also defined migration in wood turtles *Clemmys insculpta* by the synchrony of their individual movements among different activity areas between years. In Canada, at the northern limit of their range, snapping turtles *Chelydra serpentina* congregate and show strong fidelity to particular overwintering sites (Brown and Brooks 1994). Many semiaquatic species and aquatic species inhabiting seasonal and ephemeral water bodies shift their activity areas spatially and temporally according to environmental conditions. In south-western Australia, adult *Pseudemydura umbrina* tend to use the same aestivation sites every summer, and about one third of a particular population congregates in one underground tunnel system. A *Chelodina longicollis* population in coastal eastern Australia uses, during wet periods, highly productive, ephemeral swamps where they grow and reproduce; during droughts and dry years, the turtles move several kilometres overland to seek refuge in two permanent dune lakes where reproduction more or less ceases (Kennett and Georges 1990).

Migratory events of terrestrial tortoises fall mainly into the category of moving within familiar areas, but the movements of Galapagos giant tortoises *Geochelone nigra* from the cooler, moist highlands to the coastal lowlands to lay eggs can also be classified as breeding migration; at some places, migrating tortoises have worn out single-file passage ways through lava cliffs (Van Denburgh 1914; Townsend 1925). Part of a giant tortoise (*Geochelone gigantea*) population on Aldabra Atoll moves or migrates seasonally between the coast and the inland in response to changes in food availability, interacting with its food supply similarly to other large, migrating herbivores (Gibson and Hamilton 1983). Adult migrating individuals differ morphometrically from the relatively sedentary adults, the former being longer and narrower (Swingland et al. 1989). Migrant tortoises gain from the higher productivity of the coastal vegetation at the beginning of the rains and the reproductive output of females on the coast is higher than those that remain inland, but due to a lack of shade and heat exposure mortality is also higher on the coast than it is inland (Swingland and Lessells 1979). In the northern part of their range desert tortoises *Gopherus agassizii* congregate in late autumn in communal winter dens from which they disperse again in early spring, each individual migrating to a summer home range (Woodbury and Hardy 1948).

4.2 Sexual Behaviour

The main phases of the sexual behaviour of chelonians, which are part of their social behavioural repertoire, are finding a partner, courtship and copulation. Social behaviour essentially encompasses all aspects of chelonians' lives where agonistic behaviour and/or aggregations occur, including feeding, foraging, basking, emigration, migration, shelter-associated activities, movements, hibernation and aestivation (Harless 1979).

4.2.1 Timing of Mating

In most aquatic and many other chelonians that live secretively it is difficult and time-consuming to observe matings in wild populations. For most chelonians, tropical ones in particular, quantitative data on mating activity are lacking. Many reports of matings are anecdotal, and even if mating seasons were reported for a species it does not necessarily mean that mating may not also occur at other times of the year. For most chelonians, our knowledge of the timing of mating is still rudimentary.

Many chelonians of temperate regions mate and copulate in spring and ovulate and nest in late spring and early summer (e.g. *Clemmys guttata*: Ernst and Zug 1994). Due to the sperm storage capacity of female chelonians (see Sect. 3.5) a correlation is not required between insemination and maximal follicular enlargement or ovulation, allowing for a high degree of freedom in the timing of mating and copulation. Several spring nesting chelonians may show an autumn as well as a spring mating period (e.g. *Gopherus polyphemus*: Butler and Hull 1996) and some chelonians, although showing peaks in spring and autumn, may mate and copulate during their whole annual activity period (e.g. *Testudo hermanni*: Kuchling 1982a). This suggests a certain independence of sexual behaviour from the male gonadal state. The North American musk turtle *Kinosternon odoratum* seems to mate mainly in autumn when testes and epididymides are enlarged, plasma testosterone levels are elevated and temperature and day length are decreasing, but also mates in spring when testes are regressed, epididymides are enlarged, plasma testosterone levels are low, and temperature and day length are increasing (Mendonça 1987b). *Mauremys caspica rivulata* in Israel begins courtship in September (the peak period of spermiogenesis), but mating peaks in winter when the testes are regressed and continues into the spring (Garsith and Sidis 1985). This indicates that *Mauremys* mates during the time of the year when temperatures are low. In south-western Australia, *Chelodina oblonga* and *Pseudemydura umbrina*, which nest in spring and early summer, also have their peak mating activity in winter. *Chelonia mydas* produces several clutches during a breeding season and the females are sexually receptive for a restricted period and copulate before the onset of their nesting season, but not in the internesting periods (Licht et al. 1979).

4.2.2 Finding a Partner

Partner finding behaviour may involve extended movement, aggregation, dominance and/or agonistic behaviour. Although home ranges definitely exist in many chelonian species, territoriality has not been demonstrated in any field study. Territorial behaviour in animals is an expression of competition for space and the spatial distribution of resources (including, for example, females). It is probable that chelonians do not patrol or defend boundaries of home ranges and, therefore, do not defend patches of resources. In many species, however, dominance relationships occur among individuals that have overlapping home ranges (Harless 1979).

The probability of two sexual partners meeting each other is greatly increased by the aggregation of individuals. In chelonians which show reproductive migrations (see Sect. 4.1), males and females congregate in the waters adjacent to the nesting beaches where they mate. In the Amazonian rain forest, the terrestrial tortoise *Geochelone carbonaria* commonly aggregates at the height of the mating season under certain fruiting trees and several males often gather around single females (Moskovits 1988). Aggregation, however, is not ubiquitous during breeding seasons and some turtles may be widely dispersed when mating activity is at its peak: *Pseudemydura umbrina* in south-western Australia mates and copulates in winter when wide areas of their ephemeral swamp habitats are flooded. During this time males move much longer distances than females and try to mate with any conspecific they encounter. When the water retreats in spring into small puddles, the turtles congregate there and gorge themselves on the concentrated food resources (aquatic invertebrates). At that time, however, agonistic behaviour is minimal and mating ceases completely.

To find the right partner it is important to recognise one's own species as well as the opposite sex. Odour production and discrimination plays a major role in partner and possibly also species recognition of many chelonians. Smelling takes place in many encounters. However, odour and olfactory stimuli have not been investigated in detail. All chelonian families with the exception of the Testudinidae have two pairs of glands located between the anterior and the posterior edges of the plastral bridges which, in some species, exude a liquid with a strong musky odour when animals are distressed. This may be a protective device to repel predators. The function of the odour as an alarm signal or for other intraspecific communication, however, has not been thoroughly investigated (Manton 1979). In desert tortoises (*Gopherus agassizii*), chin gland secretion seems to play a role in partner or competitor recognition during the mating season (Bulova 1997). During social encounters, smelling is also often directed to the tail or the cloaca, and odour other than that of the glands near the bridges may also be important. In social interactions, olfactory discrimination may be more basic and primitive than visual discrimination (Harless 1979).

Some of the morphological differences which humans use to distinguish species must also be visual cues used by chelonians themselves for species discrimination. Differences in postures and movements may also serve as visual stimuli. The shell and/or the skin of the head, neck and limbs of various chelonians is colourful or strikingly patterned, and may facilitate species as well as sex recognition.

A number of chelonians have been described as sexually dichromatic. Cooper and Greenberg (1992) list 37 species in nine families as sexually dichromatic and nine of those as seasonally dichromatic. Melanism is a common phenomenon in males of many populations of *Trachemys scripta*, but also occurs in a variety of other turtle species. In a few species, both sexes may become melanistic. Lovich et al. (1990) distinguished between permanent melanism (species in which both sexes are black throughout life), seasonal melanism (see below), senile reticulate melanism which affects only the oldest individuals in some species, and ontogenetic melanism. In the last case the development of the melanic condition is progressive, generally occurs in males and is associated with age or body size. The few chelonians that undergo colour changes during the breeding season mainly show changes in the head and/or soft skin coloration, but in some the shell may also become brighter. Males of the tutong *Batagur baska* develop black head and shell pigmentation during the breeding season (a case of seasonal melanism), whereas males of the closely related *Callagur borneoensis* become lighter and develop a contrasting pattern on their head, including a bright red head stripe (Moll et al. 1981). In *Geochelone forstenii* and *G. elongata*, the skin around the eyes and nostrils of both sexes turns from pink to red (Auffenberg 1964; Pritchard 1979a). In those chelonians, as well as in *Kachuga trivittata* (Moll et al. 1981), the red coloration results from increased vascularisation of the skin and blood haemoglobin shining through. The seasonal dichromatism seems to be under hormonal control. Interestingly, the colourful males of *Callagur* show no tendency towards territoriality or increased intrasexual aggression (Moll et al. 1981), which indicates a purely intersexual function in mate recognition.

4.2.3 Courtship and Copulation

Reports of courtship and mating of chelonians have been reviewed by Carpenter and Ferguson (1977), Harless (1979) and Fritz (1991), and quantitative analyses of chelonian courtships were reviewed by Bels and Crama (1994). The appearance of certain act systems during courtship may be seen in males of a number of genera belonging to different families, for example head bobbing, head swaying, fanning forefeet movements or stroking the female, nudging and gentle biting or nipping with the head and jaws, trailing, and cloacal touching or sniffing. During mounting for copulation, the male assumes a superior position to the female. Males of aquatic chelonians may clasp the female's shell with all four feet (or claws) or, according to species and the relative size of the male, only the hind feet. A variety of movements occur for bringing the tails together for intromission. While visual and olfactory stimuli are important in initial approach, discrimination and the premounting courtship phase, tactile stimuli have increasing significance in the final stages of mating behaviour.

Bels and Crama (1994) divided courtships in chelonians into three types: (1) the mounting-courtship type, in which most of the behavioural action patterns are performed during the mounting, (2) the premounting-courtship type, in which the male performs several behaviours before mounting, and (3) the intermediate courtship type. These courtship types are not related to the phylogenetic divergence of chelonians into Cryptodira and Pleurodira and may occur in various families.

A high diversity of elaborate premounting courtship occurs in North American turtles of the subtribe Nectemydina (Emydidae; genera *Chrysemys*, *Graptemys*, *Malaclemys*, *Pseudemys*, *Trachemys*), with a common behavioural character in this group being vibration (or titillation) of the front foot claws of the male towards the head of the female. Males of the northern subspecies of *Trachemys scripta* are smaller than females, have greatly elongated fore claws and titillate towards the female during courtship while facing her. Males of the central and South American subspecies of *T. scripta* are about the same size as females, do not show elongated fore claws and are reported not to titillate with their fore claws during courtship. The northern subspecies occur sympatrically with other nectemydins with striped heads, suggesting that the elongated fore claws and the titillation of the female during courtship are precopulatory isolating mechanisms that have been selected for in the zone of sympatry where many striped species occur (Legler 1990).

Titillation courtship behaviour tends to be stereotypic within species, but subtle differences in claw vibration and head movements may occur between species. Fritz (1991) used claw vibration as synapomorphic character in a systematic analysis of the Nectemydina, and found that claw vibration from above (with the male swimming above the female) is a shared derived character of all species of the genus *Pseudemys*. In other genera the male vibrates the claws while facing the female from a position opposite to her. The different courtship modes of the Nectemydina evolved from a common ancestral behaviour pattern of head bobbing plus claw vibration and manifold variations of this basic pattern developed by character displacement, with the highest diversity occurring in the southeastern USA where the highest species diversity is found. The stereotypic mode of foreclaw titillation patterns may have been an important step in maintaining reproductive isolation in an area of frequent sympatry (Seidel and Fritz 1997). Head bobbing in a vertical plane is widely distributed in chelonian courtship and a plesiomorphic behavioural character (Fritz 1991).

The premounting courtship pattern is mainly found in the ecological group of aquatic swimmers, whereas semiaquatic swimmers, bottom-walking and terrestrial species often use the mounting-courtship or the intermediate courtship type. Berry and Shine (1980) predicted that female choice is important in the aquatic swimmers and forcible insemination in the other group. The female role in courtship remains unstudied for most species. Forcible insemination has not been clearly demonstrated in any species and may be an unlikely concept (see Sect. 4.2.4 for discussion). Neck and head movements during courtship of many male chelonians, including of those showing the mounting-courtship type, may be used by both sexual partners for synchronising other motor patterns such as cloacal gaping, penal erection and intromission. This synchrony during mating is certainly a crucial factor in the apposition of the penial specialised structural features, which is presumably a prerequisite for copulation success (Bels and Crama 1994).

In terrestrial tortoises, acoustic communication channels are also implicated during courtship. The most common vocalisations are rhythmic grunts or bellows associated with abrupt exhalation during pelvic thrusting by mating male tortoises. *Manouria emys* also uses, in addition to the male mating vocalisation with a duration of typically less than a second, a range of separate and distinctive male and female calls during the early stages of courtship, usually associated with a head bobbing phase. During face-to-face head bobbing males emit low frequency, modulated moans of relatively long duration (continuous for 10–15 seconds) and females occasionally vocalise for a shorter duration and with greater frequency modulation (McKeown et al. 1990).

4.2.4 Sexual Selection and Sexual Size Dimorphism

Some chelonian species show impressive size differences between adult males and females. Depending on the species, either sex can be the larger one. The fact that some species show no dramatic size difference between the sexes, whereas others have much larger females than males, and still others have larger males than females, led to various attempts to explain what determines adult body size. It can be assumed that sexual selection plays a major role in generating size differences between the sexes. According to Darwin (1871), sexual selection differs from natural selection and operates on individual characteristics of a particular sex to enhance an individual's probability of success, relative to other members of the same sex, in some aspects of reproduction. On the basis of the trait on which selection operates, sexual selection can be subdivided functionally into two categories: (1) intrasexual selection, acting on traits which provide an individual with a competitive advantage in interactions with the same sex, and (2) intersexual selection which acts on traits that make an individual more likely to be chosen by a member of the other sex, for example female choice of a male for mating.

Berry and Shine (1980) inferred from literature data that patterns of chelonian sexual size dimorphism correlate with habitat type and male-mating strategy and suggested the following scenario. (1) In most terrestrial species, males engage in combat with each other; males typically grow larger than females. (2) In semiaquatic and "bottom-walking" aquatic species, male combat is less common, but males often forcibly inseminate females; as in terrestrial species, males are usually larger than females. (3) In truly aquatic species, male combat and forcible insemination are rare; instead, males utilise elaborate precoital displays, and female choice is highly important; males are usually smaller than females.

Berry and Shine's (1980) predictions about how sexual selection operates in chelonians are, however, rather speculative and may be oversimplified. For example, the notion of the occurrence of forcible insemination is speculative and still unproven. In fact, in all chelonian species the female ultimately determines whether copulation occurs because she must relax her tail and cloaca and, in some species, elevate her rear carapace or relax her rear plastron to allow intromission. In many species male turtles may not be able to achieve intromission with recalcitrant females. Gibbons and Lovich (1990) concluded that forcible insemination as a requirement for procreation in turtles is not a believable concept.

Gibbons and Lovich (1990) presented a model for the evolution of sexual size dimorphism in chelonians in which natural selection and sexual selection operate (1) independently and in concert or (2) in opposition to produce the ultimate species-specific size difference of the sexes. Critical life history traits on which the two forms of selection operate are age at maturity, size at maturity and continued growth after maturity. Larger size in females can result in more or larger eggs, whereas larger size in males may result in superiority in male–male encounters, the potential for moving greater distances in search of new mates, and possibly an advantage as a consequence of female choice for larger males. On the other hand, either sex could benefit from maturity at a young age because of the competitive advantage of entering the breeding population early, thus potentially increasing the number of mating opportunities and offspring in a lifetime. For example, a male turtle is in competition with others in the timing of its entry into the breeding population. Becoming a competitive breeder at an earlier age offers a sexual selective advantage. This is opposed by natural selection, however, because turtles reaching maturity at a younger age grow more slowly than those remaining immature, and larger size would reduce predation. Therefore, natural selection should operate against attainment of maturity at a young age and small size.

Complications in quantifying the degree of sexual size dimorphism include the high variability of age and size at maturity among populations and continuing changes in body size due to indeterminate growth of the adults in many chelonians, meaning, for example, that the age structure of a population will affect measurements of sexual size dimorphism. With an elaborate quantification of sexual size dimorphism and use of a sexual dimorphism index, Gibbons and Lovich (1990) concluded that there is an overall trend toward larger females and that a relationship also exists between sexual size dimorphism and shell shape, but there is no relationship between species, body size and sexual dimorphism. They found a preponderance of species in which females are larger than males and that, in the case of species with larger males, the males never attain the size advantage relative to females that females of other species attain relative to males. According to their model, the smaller sex matures at a smaller size and younger age than the larger sex, and this differential size and age at maturity corresponds to the ultimate difference in size between the two sexes. However, this theory may not be ubiquitous: contrary to these predictions, the earlier maturing males in an insular population of the Florida box turtle *Terrapene carolina bauri* were found to grow to larger sizes than the later maturing females (Dodd 1997).

In terrestrial habitats and situations where large size is important to avoid predation or any other environmentally induced mortality (e.g. through desiccation or thermal stress), males of many species are the same size as or larger than females. To combat predation, larger size also operates interactively with domedness. Natural selection will favour a much larger minimum size at maturity in terrestrial chelonians than in aquatic ones, thus overriding the pressure of sexual selection for male turtles to mature at a younger age and smaller size. The phenomenon that males grow larger than females in some species, mainly the snapping turtles (Chelydridae) and many tortoises (Testudinidae), is based on sexual selection, due to the advantage of larger size in combat situations and in outcompeting other males for access to mates (Gibbons and Lovich 1990).

Yasukawa et al. (1996), in analysing sexual size differences in different populations and subspecies of *Mauremys mutica*, hypothesised again that the larger size of males in *M. mutica kami* may have most likely evolved through epigamic selection involving forcible copulatory behaviour. The authors did not clarify whether the apparently forcible copulation by male *M. mutica* is really "forcible" or whether an aspect of female mate choice is also involved or necessary. In order to test the different hypotheses, it will be essential to examine in more detail and in various chelonians the correlations between patterns of sexual size dimorphism and copulatory behaviour.

Fitch (1981) documented a trend for female reptiles to be larger, relative to conspecific males, in temperate-zone species than in tropical species and interpreted this pattern as an adaptation to the low reproductive frequencies of temperate-zone species. Females experiencing long intervals between reproductive events thus would have a selective advantage through larger body size and larger per-clutch fecundity. Forsman and Shine (1995) tested the hypothesis that the frequency of reproduction has a relationship to sexual size dimorphism in freshwater turtles (Emydidae) and found that, in contradiction to Fitch's hypothesis, the degree of sexual size dimorphism in favour of females increased with increasing annual clutch frequency across species. In turtles, the same factors that impose restrictions on the number of clutches that can be produced per year may set limits also on turtle growth rate, body size and survival. On the other hand, in an evolutionary sense, observed sexual size dimorphism may not necessarily be an expression of females growing larger, but rather of males maturing earlier and remaining smaller. High reproductive frequency may impose stronger selection for early maturity in males because of the greater potential increase in the number of matings that can be obtained in a lifetime.

4.3 Nesting

Chelonian nesting occurs in a well-defined sequence of steps that has been behaviourally characterised and interspecifically compared. Commonly distinguished phases are: nest site selection; preparation of the nest site; digging the nest hole (egg cavity); oviposition; and filling in and covering the nest (Ehrenfeld 1979; Hailmann and Elowson 1992; Kuchling 1993b). In the case of aquatic turtles, nesting generally occurs on dry land and involves the movement from the aquatic to the terrestrial habitat. In sea turtles, the approach to the nesting habitat (beach) and the departure from the nesting area are sometimes included in the phases of nesting (Hailmann and Elowson 1992).

Most chelonians construct a nest on dry ground by digging a hole in the soil or sand, let the eggs which emerge from the cloaca drop into the hole which is then filled in by the material scooped out during digging. Once the area has been disguised the nest is usually abandoned by the female (but see Sect. 4.3.7). Egg environment and incubation conditions depend, therefore, on the selection of the nest site and the mode of nest construction by the female.

4.3.1 Nesting Area and Selection of Nest Site

Chelonians occupy diverse climatic zones and habitats and it is not surprising that the criteria for nesting areas and nest site selection are broad and the associated behaviours flexible. The phenomenon that some species or population migrate to particular nesting areas has already been discussed in Section 4.1.

Since, in most chelonians, the maximum depth of the egg cavity is determined by the length of the fully extended hind legs, after tilting of the shell is taken into account (Ehrenfeld 1979), large-bodied species typically construct deeper nests than smaller ones. Many large-bodied chelonians prefer relatively open and sunny nesting areas and nest sites, like sand beaches in the case of sea turtles and sand banks in the case of large freshwater turtles, or open, hot lowlands with fine soil in the case of Galapagos tortoises. The eggs of these large species are buried deep enough to smooth the temperature extremes of the bare, sun-exposed ground. In hot climates, many smaller chelonians nest in areas with some ground cover, although temperate-zone species seem to select sites with high sun exposure and little vegetation cover.

Regarding the type of soil, the criteria for suitable sites are surprisingly broad, both among and within species. Where a preference is shown, it is usually for sandy and other friable, well-drained soils rather than clay and hard-packed soils. Several emydid and kinosternid turtles in the southern USA may construct their nests in the nesting mounds of alligators (Ehrenfeld 1979).

Some species of the families Chelidae, Kinosternidae, Dermatemidae and Testudinidae may select unusual nest sites and also depart from the typical pattern of nest construction. *Kinosternon odoratum* and *K. minor* in the southern USA and *Platemys platycephala*, *Phrynops gibbus* (Ernst and Barbour 1989) and *Geochelone denticulata* (Ehrenfeld 1979) in the Amazon area often deposit their eggs on the ground beneath leaf litter or beneath the edge of a fallen log without digging nest holes in the ground. Poorly covered nests are typically in habitats that are shadier and more moist than the sites of covered nests.

Chelodina rugosa in northern Australia nests in flooded plains under water (Kennett et al. 1993), and the tropical *Dermatemys mawii* nests in riparian vegetation or at clay banks which are temporarily submerged (Polisar 1996). The nests of these species are constructed in the typical manner and to typical depths, with the eggs only progressing with development when the nests fall dry.

4.3.2 Conditions for Nesting

Chelonian species prefer to nest either during the day or at night. Sea turtles nest generally during the night while most other species nest during the day. In addition to daylight, Australian chelids seek out rainy weather conditions to nest (Goode 1965, 1967; Clay 1981; Kuchling 1993b). Most of them continue nesting during the night if they cannot finish nesting in daylight (e.g. *Emydura macquarii*: Goode 1965; *Chelodina expansa*: Georges 1984; *Chelodina oblonga*: Burbidge 1967). *Pseudemydura umbrina* is the only chelonian known to sleep in the unfinished nest when it cannot finish nesting before dusk and to complete nesting the following morning (Kuchling 1993b).

4.3.3 Preparation of Nest Site

Typically, in the species that prepare the nest site, the forefeet are used to scratch, scrape or throw soil substrate or overlaying material backward or to the side. Preparations of more or less developed "body pits" have been found in marine, freshwater and terrestrial chelonians, but many species do very little preparing of the nest site before digging an egg cavity.

The large tortoise *Manouria emys*, which lives in forests of Southeast Asia, constructs a nest mound by back-sweeping leaf litter and ground material with the forelimbs; the nest is then dug into this mound (McKeown et al. 1990). The mud turtle *Kinosternon flavescens* nests while completely buried underground in the sandy soil of sand hills; females dig down to an average soil depth of circa 13 cm (to the top of the carapace) by pushing the sandy soil laterally with the action of the forelimbs and posteriorly with the hind limbs; while buried horizontally at this depth, and completely surrounded by soil, the female excavates her nest (Iverson 1990). This can be interpreted as modified nest site preparation, which allows the relatively small species to deposit the eggs deeper in the soil than the conventional nesting pattern would allow.

4.3.4 Digging the Nest Hole

Digging the nest hole or egg cavity is one of the most stereotypical behaviours of chelonians as a group. The nest hole of most species is flask-shaped, wider at the bottom than at the top, the result of an exceptional stereotyping and replication of the movements of the hind legs and the body. Until recently it was commonly accepted that all chelonians dig the nest cavity exclusively with the hind feet. "We do not know what evolutionary forces prompt a gopher tortoise, *Gopherus polyphemus*, which is capable of digging a 30 ft burrow with its fore feet, to dig an egg cavity with its elephantine hind legs, removing scarcely a few grains of dirt with each stroke; but these forces must be conservative indeed" (Ehrenfeld 1979: 422).

The only published exception to the rule that chelonians dig the nest hole with their hind feet is *Pseudemydura umbrina* which digs the nest hole with the forefeet (Kuchling 1993b). *Pseudemydura umbrina* is the smallest chelid species of Australia. Females start laying eggs with 113 mm carapace length and reach a maximal carapace length of 135 mm (Kuchling and Bradshaw 1993). Eggs are laid in late spring, in November and early December, when the ephemeral swamps which the species inhabits dry out; the eggs remain in the nest over the hot, dry summer months until hatchlings emerge, after about 180 days, during May and June (Burbidge 1981). The harsh environment, together with the stringent anatomical limitation of the short length of the hind legs due to the small size of *P. umbrina* females, may be the reasons for its alternative, non-conservative way to construct the nest. In order to facilitate successful development of its eggs, *P. umbrina* may have to bury the eggs deeper into the soil than the conservative stereotypical nesting behaviour of chelonians would allow. *Kinosternon flavescens,* which nests while completely buried underground (Iverson 1990), obviously solved the same problem with another strategy.

4.3.5 Oviposition

All chelonians which dig the nest hole with the hind feet complete it without any visual inspection before commencing oviposition. Tactile feedback from the hind legs and the tail may tell the female when the nest is ready. In species with a narrow posterior shell opening and relatively large eggs, the plastron and carapace may have to move apart to allow passage of the eggs (Fig. 4.1). Sea turtles simply let the eggs drop down into the hole, but other chelonians, including testudinids with their elephantine feet, commonly use their hind feet to position eggs in the nest cavity. Among species producing small clutches, all of the eggs eventually have contact with the substrate. Among species producing large clutches, many eggs are supported by points of contact with other eggs and make no contact with the substrate.

4.3.6 Covering the Nest

Immediately following oviposition, chelonians start filling in the nest hole with their hind legs. The material which was dug out of the hole is pushed and scooped back in. During the process of filling in the nest, some soil material sifts downward among the eggs, but, depending on the species-specific shape of the nest

Fig. 4.1. *Pseudemydura umbrina.* Oxytocin-induced oviposition under water; despite the fact that, in all side-necked turtles (Pleurodira), the pelvis is rigidly attached to the carapace as well as the plastron, carapace and plastron of *P. umbrina* move apart during oviposition to allow the passage of eggs through the narrow posterior shell opening. The mechanism which allows this kinesis is unknown

hole and the substrate, an air space may remain between the eggs and the top of the cavity. In stable soils these air spaces persist for the duration of incubation, but they may be obliterated by infiltration of soil into nests located in less stable substrates. In some conditions, persistence of air space may be important for successful incubation (Packard and Packard 1988).

Chelonians often compact the loose soil above the nest. *Podocnemis unifilis* packs the sand by stamping with the hind legs (Foote 1978) and *Emydoidea blandingii* packs the nest by kneading the soil with the hind feet (Linck et al. 1989). Various species also use their plastron to smooth and/or compact the nest site. Freshwater turtles of different families may raise themselves up on all four legs and drop directly down on the nest. Sea turtles that dig body pits use all four legs to fill them in, gradually moving forward and also disturbing areas adjacent to the nest. In *Pseudemydura umbrina*, due to its mode of nest construction with the forelegs, the nest entry is not directly above the eggs, but about 10 cm to one side (Fig. 4.2). Since the female starts nesting at the margin of a grass tussock, leaf litter or some other structure, the area directly above the eggs has not been disturbed and typically has some cover above the soil surface. In most cases no surrounding material other than earth was swept over the nest, but, with some nests in areas of grass, surrounding grass was rearranged over the entry.

4.3.7 Nest Guarding

Nest guarding is a form of parental care which is uncommon in chelonians, but also little studied. *Manouria emys*, a large Asian forest tortoise which constructs

Fig. 4.2. Longitudinal section of a finished nest of *Pseudemydura umbrina*. Eggs (one is visible) are 70–90 mm underground; entry of the nest is about 100 mm sidewards of the eggs. Ground above the eggs is undisturbed and densely covered with vegetation. *Dashed line* indicates margins of the filled-in tunnel; scale in centimetres

a nest mound (see Sect. 4.3.3), shows the most elaborate nest guarding behaviour known for any turtle. After oviposition and filling in the nest cavity, the female continues to increase the size of the nest mound and, over several days, adds to and shapes the nest mound by collecting additional litter from adjacent areas or even by biting off clumps of grass for this purpose. Captive females show aggressive nest guarding behaviour for 3 to 20 days after oviposition. If a human (or a stuffed monitor lizard) approaches a nest, the female either moves to a position on top of the mound, or advances towards the intruder and, with the front of her shell and the head, attempts to push the intruder away from the nest (McKeown et al. 1990). Wild female desert tortoises *Gopherus agassizii* were also observed aggressively defending by pushing and blocking nests and eggs from Gila monsters *Heloderma suspectum* (Barrett and Humphery 1986).

Female yellow mud turtles (*Kinosternon flavescens*), which nest while completely buried underground, remain buried directly above their egg clutch for from 1 to more than 38 days. This nest attendance seems to be inversely correlated with rainfall and may reduce egg mortality by the effects of the female's presence on both predation of the eggs and soil moisture at the nest (Iverson 1990).

4.4 Emergence from the Nest

Apart from the eggs, the hatchlings are the most vulnerable stage in the life of chelonians. Emergence from the nest is a critical phase in particular for aquatic species, since hatchlings have to move overland from the nest to the aquatic habitat, which exposes them to terrestrial predators and, potentially, to temperature extremes and desiccation. Nest emergence is well studied in sea turtles and a few other colonial breeders, but few data are available for other groups.

The egg chamber of sea turtles is about 1 m down in the sand of the sea beach. After hatching, the hatchlings are confined to this subterranean chamber where their carapaces flatten and the external yolks are resorbed as an energy store. The discarded egg shells are shuffled downwards in the egg chamber. The process of climbing to the surface involves collective thrashing, which shifts the loosened sand down between the hatchlings to the floor. The mutual stimulation by the hatchlings was described as "social facilitation" by Carr and Hirth (1961). The outbursts of thrashing are sporadic, usually triggered by one turtle and quickly spreading through the clutch. The periods of quiescence may be obligatory when the oxygen demand exceeds the rate of oxygen diffusion into the chamber, or may be due to temperature inhibition. This process moves the hatchlings, and their chamber, upwards until the dry surface sand caves in and forms a crater. The movement to the beach surface takes 3 to 7 days or longer if the sand surface is packed. The hatchlings, then, remain quiet for a while, evidently awaiting the temperature drop that usually occurs during the night. Emergence usually occurs *en masse* at night, but it may take up to 3 days for the entire clutch to emerge (Miller 1985, 1997; Carr 1995).

The nocturnal emergence of sea turtle hatchlings is controlled by the temperature gradient in the sand. As hatchlings dig towards the surface during the day, they reach warmer sand. While the sand above them is warmer, digging is inhibited. At and after sunset the surface temperature of the sand drops rapidly. The latent heat contained in deeper sand layers reverses the temperature gradient. When the temperature of the sand at the level of the hatchlings has dropped the hatchlings dig again towards the surface, arriving after dark. Hatchlings may, however, also emerge during the day on cool, cloudy and/or rainy days (Lohmann et al. 1997; Miller 1997).

After emergence from the sand the hatchlings crawl as quickly as possible towards the sea and into the surf to avoid terrestrial predators like mammals, birds and crabs. Hatchlings of all marine turtles are able to take accurate seaward headings from most nest sites even when the surf is not in view. Light is clearly important in the guidance process. Obviously, light angle and intensity differences that are measured at the retina initiate the turning toward brightness. Hatchlings seem to average brightness input over a wide horizontal range, about 180 degrees, and a narrow vertical range of about 10 degrees. Under natural conditions, hatchlings find the sea by crawling toward the brighter, lower oceanic horizon and away from the elevated silhouettes of vegetation and dunes of the landward edge of the beach. Although hatchlings tend to orient away from vertical stripes and elevated silhouettes irrespective of the brightest direction, highly directed light fields of artificial light sources near nesting beaches may become supernormal stimuli, misdirect hatchlings and cause substantial mortality (Witherington 1995).

Wavelength of the light is also important, and behavioural responses to specific wavelengths differ among species. Green turtle (*Chelonia mydas*), hawksbill (*Eretmochelys imbricata*) and olive ridley hatchlings (*Lepidochelys olivacea*) are strongly attracted to light in the near-ultraviolet to yellow region of the spectrum (360–600 nm), whereas loggerhead (*Caretta caretta*) hatchlings show an aversion to yellow to green–yellow light (560–600 nm). The threshold intensity of the light which is required to elicit a response differs among wavelengths. Ultraviolet–green light requires the lowest intensity, the threshold for yellow light is about one order of magnitude higher in all turtle species and that for red light about five orders of magnitude higher. The adaptive significance of these responses may be to factor out light from the sun or moon near the horizon: a rising or setting sun or moon is rich in yellow light and, indeed, alters little or not at all the sea-finding ability of the hatchlings (Lohmann et al. 1997).

As soon as the hatchlings encounter water, typically the shallow sheet flow of a spent wave, the crawling locomotion is replaced by wing-like swimming strokes. The seaward bearing is maintained by positive rheotaxis and when the backwash starts sliding seaward the polarity of the current cue must reverse because the heading of the hatchlings remains the same. The hatchlings avoid breaking waves by diving to the bottom and riding the undertow, reappearing on the surface in the relatively calm water beyond (Carr 1995). Nearshore aquatic predators, which include numerous fish, sharks and invertebrates, may cause even heavier losses of hatchlings than the terrestrial predators (Stancyk 1995) and hatchlings keep heading out to the relative safety of the open sea (see migration, Sect. 4.1).

Nests of the arrau turtle *Podocnemis expansa*, which nests communally on sand banks, show at least two hatching waves within one clutch. First about 60% of the eggs hatch and the turtles move up in the sand to a level 20 cm from the surface. About 3 days later, the remaining eggs hatch and move to join the first group. The hatchlings leave the nest cavity at night, especially rainy nights, and run directly to the river (Alho and Pádua 1982).

Many chelonian hatchlings need an environmental trigger to emerge from the nest. In sea turtles the trigger seems to be the temperature gradient in the top 10 cm of the sand column (Gyuris 1993). Sea turtles and many river turtles which nest in rather unstable sandy environments have short incubation periods (Ewert 1979) and, typically, emerge without major delay. The hatchlings of many other chelonians, however, may remain dormant in the nest for prolonged periods before emergence. The adaptive significance of the variation in the timing of emergence is best understood in terms of survival probabilities of the hatchlings (Gibbons and Nelson 1978). Tropical tortoises and turtles often emerge during the rainy season, with emergence being associated with the onset of heavy rainfall and/or flooding of nests. The hatchlings of several freshwater turtles of temperate zones may overwinter in the nest. For them, cues may be the spring warm-up which follows the winter cold. In south-western Australia, *Chelodina oblonga* hatchlings typically emerge in early spring after 6–11 months of incubation (Clay 1981), whereas, in the same area, hatchlings of *Pseudemydura umbrina* emerge in late autumn. In *P. umbrina*, hatching of eggs can be induced by a drop in incubation temperature (Burbidge and Kuchling 1994), but hatchlings only emerge from nests after heavy rains. Hatchlings of nests in clay soil emerge considerably earlier in the season than those of nests in sandy soil, suggesting that low oxygen or high carbon dioxide concentration may be the trigger for emergence rather than moisture per se. *Pseudemydura umbrina* inhabits ephemeral swamps, and hatchlings have to exploit any standing water to grow enough during their first winter and spring so they can survive the next dry summer. *Chelodina oblonga* inhabits permanent or semi-permanent waters where other survival factors may operate.

In continental areas of the northernmost distribution of chelonians in North America, where winters are cold and harsh, hatchlings of different species apply different strategies to overwinter. Hatchlings of *Chelydra serpentina* and *Emydoidea blandingii* emerge from the nests in late summer and move into lakes and ponds to spend the winter in the water. Hatchlings of *Kinosternon flavescens* and *Terrapene ornata* dig down through the bottom of their nests and spend their first winter deep in the soil to avoid freezing. Hatchlings of *Chrysemys picta* remain inside their shallow nests in the frozen soil throughout the winter and do not emerge until the soil thaws the following spring (Packard and Packard 1997). Hatchlings of *C. picta* overwintering in nests survived temperatures as low as $-12.7\,^{\circ}C$, which is lower than the lowest body temperature that has been reliably reported to be tolerated by any other vertebrate (Packard et al. 1997a). Three hypotheses have been proposed to account for the tolerance of hatchling *C. picta* to cold. Storey et al. (1988) and Churchill and Storey (1992) proposed that the turtles tolerate freezing. Costanzo et al. (1995) suggested that overwintering *C. picta* hatchlings use supercooling, but commonly survive brief, shallow freezing when they are unable to avoid inoculation. However, Packard et al. (1997b) recently demonstrated with a new experimental protocol in the laboratory that *C. picta* hatchlings survived extended periods at subzero temperatures up to $-10.3\,°C$ unfrozen in a supercooled state, but succumbed as soon as their body water froze (which in some cases even occurred at a temperature of $-2.3\,°C$). Thus, earlier reports of freezing tolerance were not corroborated by more sophisticated experiments. *Chrysemys picta* hatchlings survive by supercooling due to a cutaneous barrier that impedes the penetration of ice into body compartments (inoculation) from frozen soil. Neonates of the other species do not have this barrier and must avoid contact with ice by digging out of the nest cavity before the arrival of cold weather (Packard and Packard 1997).

Few investigations have been reported of the behaviour of neonatal freshwater turtles once they have left the nest. For turtle species that nest close to water, terrestrial post-emergent behaviour of hatchlings may be brief and have little influence on recruitment, but, in species which commonly nest far from water, their behaviour may play a significant role for survival and recruitment into the population. Butler and Graham (1995) observed that *Emydoidea blandingii* hatchlings use olfactory cues in search of wetlands and that, on their way, they follow scent-trails of other hatchlings. During their overland movements, hatchlings burrowed into cryptic forms among mosses, leaf litter and grass tussocks, both at night and during diurnal temperature extremes. They moved predominantly in the early and mid-morning and in the late afternoon. In *Malaclemys terrapin*, which inhabits tidal marshes of the USA Atlantic coast, hatchlings emerge from the nest during day time and move towards the closest terrestrial vegetation – an apparently negative phototaxis (Burger 1976). If released in the water, *Malaclemys* hatchlings display a general avoidance reaction to open water, swim toward shoreline vegetation and burrow through dense mats of vegetation at the high tide line (Lovich et al. 1991). Cryptic behaviour may be the main reason for infrequent encounters with hatchlings in population ecological studies.

After emergence in autumn, *Pseudemydura umbrina* hatchlings typically hide for some weeks under leaf litter, mosses and grass tussocks until swamps or clay pans fill with water. Hatchlings are very difficult to find, but if, during winter and spring, a hatchling is found in the water, there is a very good chance that another one to two hatchlings can be found in the immediate neighbourhood. This suggests that hatchlings may stay together in groups. The same may occur in *Chelodina oblonga*. Aquatic drift fences with unbaited funnel traps trap hatchlings and small juveniles only sporadically, but nearly always in groups of 2–7 in one trap (G. Kuchling, unpubl.).

Resource needs may be different in hatchling and juvenile aquatic turtles which often have a more carnivorous diet than adults (Parmenter and Avery 1990). Tortoises and terrestrial turtles (for example box turtles of the genus *Terrapene*) generally have relatively larger eggs and, therefore, larger hatchlings than aquatic species with similar body size. Hatchlings of aquatic species move to highly productive habitats containing high food densities, whereas hatchlings of terrestrial species emerge into a harsher and more variable environment in which the distribution of food is less dense and probably more clumped. Terrestrial chelonians are also more herbivorous and less able to attain a positive energy balance in as short a time as aquatic hatchlings. The body condition of hatchling spur-thighed tortoises, *Testudo graeca*, decreased steadily in autumn in the first few months after hatching. At the beginning of the new activity season in spring, when the water and protein content of food plants was higher than in autumn, energy was primarily allocated to building up body reserves rather than to carapace growth, and the spring growth burst of hatchlings began approximately 1 month after the start of the activity season (Keller et al. 1997). Autumn activity in *T. graeca* hatchlings possibly reflects a trade-off among the reserves provided by the egg, energy expended in foraging and energy acquired from available food resources. Butler et al. (1995) recorded lower activity levels of hatchling *Gopherus polyphemus* during the weeks subsequent to emergence and preceding dormancy than those registered for yearlings and adults in the same period. A likely explanation is that, in the first weeks after emergence, hatchlings are mainly living on yolk reserves, which are present at hatching (see Chap. 8). Juvenile desert tortoises *Gopherus agassizii* show no increment in energy expenditure over adults, and resting metabolic rates and water influx rates are not notably different (Nagy et al. 1997). It is possible that the neonates have behaviours that reduce field energy expenditures compared with adults (less physical activity, more cryptic behaviour, less time at high body temperatures), which may mask the added costs of growth they are paying.

Chapter 5

Reproductive Cycles and Environment

In all organisms, reproduction is timed through natural selection to occur during those periods of the year that are most propitious for the survival of both parent and young. According to Baker (1938), these favourable conditions are the "ultimate cause" of a particular breeding season; the "proximate causes", which are another set of factors, initiate the sexual cycles at the appropriate time of the year (see Chap. 6). In the following account I will firstly discuss cycles according to the climatic zones, then discuss multi-year cycles and then some special adaptations to unpredictable or extreme environments. The focus of this chapter is on gonadal, mating and nesting cycles, and their timing.

Licht (1984) categorises the extent of seasonal variation of reptilian gonadal activity as follows: (1) continuous or acyclic breeding with comparable levels of reproductive activity in all months; (2) continuous breeding, but with variable levels in the intensity of reproductive activity; or (3) discontinuous or seasonal breeding with periods of gonadal activity alternating with periods of quiescence. All chelonians for which sufficient reproductive data are available seem to fit the categories (2) and (3). Acyclic breeding as defined by Licht (1984) has not been demonstrated in any species. Climatic zones influence reproductive cycles, but cycles of sympatric species often do not correspond, which indicates that evolutionary history and biogeography also play an important role in the timing of the cycles.

5.1 Temperate Zone

Winter in temperate climates is an important proximate factor for reproductive cycles, since, in the ectothermic chelonians, cool temperatures slow down physiological processes. In short, the typical temperate reproductive cycle is as follows: spermatogenesis begins in spring and is completed by early autumn; sperm is stored over the winter in the epididymides; mating occurs after emergence from winter dormancy; females ovulate in spring; oviposition takes place from late spring to early summer; and follicular growth in preparation for next year's breeding season starts in late summer or autumn. The male and female cycles are, therefore, asynchronous. Several modifications of this basal cycle can be found.

The male gonadal cycle of temperate zone chelonians is remarkably uniform. Winter is the time of germinal quiescence and spermiogenesis generally takes place during the hottest time of the year, in summer. Some flexibility occurs in the onset of gonial proliferation and spermatocytogenesis, which may start between early spring and early summer, and the duration of spermiogenesis and spermiation, which may end earlier or later in autumn. As a result, some species or populations may show a relatively short peak of spermiogenesis between summer and early autumn, whereas others may have a longer plateau of spermiogenesis and spermiation. Spermatozoa are stored in the epididymis where they may be found from the onset of spermiation onward throughout autumn, winter and spring and, sometimes, until early summer. Since most species seem to show a mating peak in spring, temperate zone chelonians exhibit a postnuptial spermatogenetic cycle.

The female reproductive cycle can be divided into the four phases of (1) follicular enlargement, (2) ovulation and intrauterine period, (3) nesting period and (4) latent period (for review see Moll 1979). A common pattern is that follicular enlargement (or vitellogenesis; see Sect. 3.2) begins in late summer or autumn and, after a break during hibernation, continues until completion in spring. Depending on geographic location, age and species, the yolk accumulation for the first clutch may be completed either before or after hibernation. In hibernating *Testudo graeca* females, liver lipids and proteins continue to be mobilised for vitellogenesis and decrease to levels ten times lower than in males (Ricceri 1953).

In the northernmost temperate zones of the chelonian distribution, the short activity period leads to variations of the standard pattern of gonadal activity. *Chelydra serpentina* of the northern temperate area (Wisconsin) shows rather compressed gonadal cycles, with vitellogenesis starting more or less immediately after nesting in mid-July and being completed by mid-September when follicles reach preovulatory size. This compression of follicular growth means that vitellogenesis in this population takes place at the same time and parallel to spermiogenesis. Ovulation occurs again in spring (May or June); mating is non-seasonal (Mahmoud and Licht 1997). The ovarian and testicular cycles of northern *C. serpentina* are shorter than those reported for other chelonians and can be regarded as synchronised. A different strategy to accommodate vitellogenesis in the short activity period of cool climates may be the extension of the growth of individual follicles over several years.

Follicles, typically, tend to enlarge in sets of distinct size groups. Agassiz (1857) suggested that these sets represent clutches to be laid in subsequent years, whereas Moll (1979) attributed them to multiple clutches that are laid in a single breeding season. Both concepts seem to be valid for different species and populations of temperate-zone turtles, but reported data for many species are still insufficient to categorise them with confidence (see Sect. 3.2.2 for discussion). Congdon and Tinkle (1982), for example, suggested that smaller size classes of vitellogenic follicles in the ovaries of *Chrysemys picta* in spring may represent either a second clutch for the next nesting season or a clutch which will be layed a year later. Ernst and Zug (1994) concluded that, in the North American *Clemmys guttata* (Emydidae), vitellogenesis of individual follicles takes more than 1 year, even though females lay eggs annually. Two or three sets of follicles of different size classes occur in all females at any time of the year, suggesting that vitellogenesis is continuous (and, in that sense, acyclic) except during winter dormancy (Ernst and Zug 1994). Moll's (1979) conclusion that the number of groups of enlarged follicles present immediately before the nesting season indicates the number of clutches to be laid in a single season does not apply for all temperate zone turtles, in particular not to species or populations in cooler northern regions.

To summarise, males of temperate zone chelonians show a postnuptial spermatogenetic cycle. Females show postnuptial vitellogenesis which is characterised by an extended period of follicular growth that commences shortly after the breeding season. (Prenuptial vitellogenesis is characterised by rapid follicular growth immediately before ovulation, with the ovaries remaining relatively quiescent during most of the year: Licht 1984.) All the patterns discussed above can be interpreted as variations of the basal temperate zone female cycle of follicular growth in late summer/autumn and early spring, oviposition in spring and hatchling emergence in late summer, autumn or, after overwintering in the nest, in early spring (see Sect. 4.4). Only a few turtles, the distribution of which just reaches temperate climates, e.g. *Deirochelys reticularia* in North America (Ernst et al. 1994) and *Chelodina expansa* in south-eastern Australia (Legler 1985), show different reproductive cycles, with nesting and egg laying taking place from autumn to late winter/early spring. This pattern can be interpreted as subtropical or tropical. These species may have evolved from ancestral tropical or subtropical stocks.

5.2 Tropical and Subtropical Zone

Seasonal changes of daylength and temperature decrease with decreasing distance from the equator. Reproductive activity is often less seasonal in tropical than in temperate chelonians and other reptiles. In some tropical and subtropical climates there are well defined dry and wet seasons, and moisture, rather than temperature, may become the critical limiting (ultimate) factor. The smallest seasonal changes may be found in equatorial rain forests. Still, water levels of rivers may show dramatic seasonal changes, for example in the Amazon and Orinoco, where some chelonians show highly seasonal reproductive cycles.

Several attempts have been made to define tropical patterns in chelonian reproduction. Moll (1979) proposed that chelonian reproductive patterns have evolved in two diametrically opposed directions, the extremes of which are characterised as follows. *Pattern I*: multiple, large clutches of relatively small eggs produced during a well-defined nesting season; communal nesting; careful construction of covered nests. *Pattern II*: small clutches of relatively large eggs; acyclic or continuous nesting; solitary nesting with no special nest area; nests poorly constructed or not even attempted. Moll (1979) further suggested that the extreme patterns are best represented in the tropics, and most temperate-zone species have patterns somewhere between. According to Moll and Moll (1990), the family Bataguridae evolved in the tropics and the majority of the species tend toward Pattern II reproduction, whereas the mainly temperate-zone Emydidae tend towards Pattern I reproduction. This trend is maintained by neotropical slider turtles (Emydidae) which migrated into the tropics rather recently from temperate North America. They have longer nesting seasons and greater reproductive potential (larger clutches) than their temperate conspecifics and show no indication to evolve in the direction taken by the Bataguridae, their closest aquatic and semiaquatic relatives. Phylogeny, obviously, plays an important role for the timing of gonadal cycles in the tropics.

Legler (1985) defined tropical and temperate reproductive patterns for the Australian family Chelidae. The tropical pattern is characterised by large rigid-shelled eggs; long incubation period; nesting during the tropical dry season; emergence of young coincident with first heavy rains and/or floods. The temperate pattern is characterised by small, flexible-shelled eggs; short incubation period; nesting during spring and summer; and emergence of young occurring before winter. Legler (1985) assigned the generic groups *Chelodina* (*expansa*) and *Elseya* (*dentata*) to the tropical pattern and *Emydura* and *Elseya* (*latisternum*) to the temperate pattern. The tropical genera restrict oviposition to the first 8 months of the year, corresponding to the tropical dry season, whereas the temperate genera utilise the last 6 months of the year for oviposition, corresponding to the southern temperate spring and summer. The temperate as well as tropical genera range over the whole latitudinal spectrum of mainland Australia, but Legler postulated that within the genera he defined these patterns do not change over wide latitudinal ranges. The ancestral stocks of the generic groups should have evolved in the tropical north or the temperate south respectively. When they dispersed into the other climatic zones, their reproductive patterns would have remained essentially unaltered. However, recent data on reproductive patterns of Australian chelid turtles suggest that Legler's (1985) hypothesis may be oversimplified. For example, *Chelodina oblonga*, which Legler considered to belong to the *C. expansa* generic group, does not exhibit the reproductive pattern postulated by Legler for that group. The reproductive pattern of *C. oblonga* (which lives in the temperate zone of south-western Australia) is rather similar to that of the *C. longicollis* group (Kuchling 1988b), the second generic group of Australian long-necked turtles defined by Legler (1985). However, *Chelodina oblonga* may warrant its own generic group (Burbidge et al. 1974) and its close alliance to the *C. longicollis* group was recently confirmed by systematic investigations (Thomson et al. 1997).

There seems to be no generalised pattern for tropical chelonians as a whole with respect to the time of the year at which breeding and nesting occurs. Table 5.1 shows the nesting periods and incubation times of some turtles in Venezuela for which enough relevant information is available. Vogt (1997a) defined five broad categories of nesting patterns in the neotropics: (1) the rainy season; (2) end of the rainy season; (3) initiation of the lowering of river depth; (4) lowest river depth; and (5) spring nesters. The high variability in the timing of reproduction suggests also a high variability of gonadal cycles and proximate timing cues.

The reproductive activities of river turtles of genus *Podocnemis* (Pelomedusidae) of tropical South America depend on the hydrological cycle. In *Podocnemis expansa*, nesting takes place when the rivers have their lowest water

Table 5.1. Breeding seasons of some turtles in Venezuela (information summarised from Pritchard and Trebbau 1984)

Species	Family	Nesting period	Incubation time (days)
Podocnemis expansa	Pelomedusidae	Feb–Mar	45
Podocnemis unifilis	Pelomedusidae	Jan–Mar	?
Podocnemis vogli	Pelomedusidae	Nov–Jan	90–147
Chelus fimbriatus	Chelidae	Oct–Nov	208
Phrynops geoffroanus	Chelidae	Mar–Apr	~150
Platemys platycephala	Chelidae	Aug–Feb	~150
Rhinoclemmys diademata	Bataguridae	All year	?
Trachemys scripta callirostris	Emydidae	Dec–Apr	~80
Trachemys scripta chichiriviche	Emydidae	Feb–May	?
Geochelone carbonaria	Testudinidae	Jul–Sep	~150
Geochelone denticulata	Testudinidae	Aug–Feb	136

levels and sand banks are exposed. Different rivers or tributaries may reach this point at different times and the turtle populations time their nesting according to this pattern. In Colombia, for example, *Podocnemis unifilis* nests between July and November on the Amazon near Letitia, but from October through February along the Putumayo (Foote 1978). In the Venezuelan llanos, *P. unifilis* nests from January to March (Pritchard and Trebbau 1984).

Sperm production is energetically less demanding than egg production, and temperatures in the tropics may allow continuous spermatogenesis. However, the few tropical species in which spermatogenesis was studied suggest that at least some cyclicity still occurs. Flores-Villela and Zug (1995) proposed that a population of *Claudius angustatus* in tropical Mexico shows continuous spermatogenesis at the population level, although individuals may have discontinuous spermatogenesis with a resting period between February and May. In that population, testes may never fully regress and spermatocytogenesis may be more or less continuous, but spermiogenesis and spermiation still peaks from June to January and is cyclic in this respect. Many subtropical species show spermatogenetic cycles which basically resemble those of temperate-zone chelonians, for example *Trachemys dorbigni* in southern Brazil (Silva et al. 1984) and *Chelodina steindachneri* in subtropical Western Australia (Kuchling 1988b).

Although data on nesting seasons are available for some tropical and subtropical chelonians, descriptions of the actual gonadal cycles are still scarce. In the Indian softshell turtle *Lissemys punctata granosa*, males as well as females show highly seasonal gonadal cycles. The male cycle corresponds to the typical temperate pattern, with a progressive phase from March to June, peak testicular activity during July/August (the summer monsoon season), regression from September to November and a quiescent phase from December to February. The female cycle parallels the male cycle, with follicular growth from February to August, oviposition in September, ovarian regression from September to November and a quiescent phase in December/January. Mating seems to occur in July/August (Singh

1974, 1977). In *L. p. granosa*, therefore, male and female gonadal cycles seem to be synchronised and prenuptial, a pattern unknown in temperate zone chelonians. The ovarian cycle of the freshwater turtle *Kachuga smithi* in Pakistan is very similar to that of *Lissemys*, with follicular growth from March to August and oviposition in September/October (Auffenberg and Khan 1991).

Sea turtles live and/or breed mainly in the tropical and subtropical zone. Males of the subtropical *Caretta caretta* show a prenuptial spermatogenetic cycle (Wibbels et al. 1990), a pattern which may apply to all sea turtles (Owens 1997). A peculiarity of the prenuptial spermatogenetic cycle of sea turtles is that sperm formation takes place during "winter" and early spring (in any case, during the cooler time of the year), which is in contrast to all other chelonians in which spermatogenesis peaks during the warmest time of the year, including *L. p. granosa* which also shows prenuptial spermatogenesis (Singh 1974, 1977). Sea turtles, males as well as females, are also special in so far as they tend not to breed on an annual basis. This phenomenon is the theme of the next section.

5.3 Multi-Year Reproductive Cycles

With the possible exception of some *Lepidochelys kempii*, female sea turtles do not reproduce every year. Most sea turtle populations exhibit more or less extensive reproductive migrations (see Sect. 4.1), which means that foraging and breeding is separated spatially as well as temporally. The mean re-migration interval of nesting females of the seven marine turtle species, calculated from the means of various populations, is: 2.9 years for *Eretmochelys imbricata* (2.74 clutches/season), 2.86 years for *Chelonia mydas* (2.93 clutches/season), 2.65 years for *Natator depressus* (2.84 clutches/season), 2.59 years for *Caretta caretta* (3.49 clutches/season), 2.28 years for *Dermochelys coriacea* (6.17 clutches/season), 1.7 years for *Lepidochelys olivacea* (2.21 clutches/season) and 1.5 years for *Lepidochelys kempii* (1.8 clutches/season) (Miller 1997). Reproductive cycles of males and females do not necessarily have the same length. For example, males of *Caretta caretta* and *Chelonia mydas* may breed every year or every 2 years. Nesting of sea turtles may be distinctly seasonal, for example during summer in subtropical eastern and western Australia, or all year round with a dry season peak in activity, as for example across tropical northern Australia (Limpus and Miller 1993).

Two aspects are interesting in the timing of gonadal cycles in sea turtles. One is the timing of the annual breeding and nesting season, which differs among regions, species and populations; the second is the period between reproductive episodes of individual turtles, the periodicity of their gonadal cycle. Some individuals may show much longer intervals between breeding events than the mean data presented above suggest. The reproductive periodicity of sea turtles is based on a cycle of energy accumulation, storage, reorganisation and utilisation, but this cycle is even less documented than gonadal cycles. Most data on gonadal cycles of sea turtles were acquired by endoscopy. With this method, it is not only possible to follow vitellogenesis (which takes 10–12 months to complete) and to predict if a female will breed in the coming season, but also to estimate retrospectively, according to the diameter of regressing corpora lutea and/or ovulation scars, if a female had bred during the previous 2 years (Limpus and Reed 1985).

A combination of the constraints of energy accumulation and of the energetic demands of a long breeding migration seems to be the most plausible explanation for the evolution of multi-year gonadal cycles in sea turtles. Once the female invests in the energetically demanding breeding migration, the expenditure of which is largely independent of fecundity, she optimises this investment by producing several large clutches of eggs and then pauses for 1 or more years to replenish energy stores. Compared to an annual production of smaller numbers of eggs plus an annual migration, the multi-year breeding strategy reduces the migration energy expenditure per egg.

There is now good evidence that, depending on the quality and quantity of food available in the foraging areas, it may take several years to accumulate the energy reserves required to support vitellogenesis and migration to the breeding and nesting areas. In the herbivorous green turtle *Chelonia mydas*, the internesting period varies according to the productivity of the foraging ground: following an El Niño–southern oscillation event which is associated with warming of sea surface temperatures and increased growth of seagrass and algae, females around northern Australia may reproduce again after 2 years, but may pause for several years during anti-El Niño–southern oscillation periods (Limpus and Nicholls 1988). Reproductive frequency does not seem to be associated with fluctuations in major weather patterns in the carnivorous or omnivorous species (Limpus and Miller 1993).

Little is known about multi-year reproductive cycles of non-marine chelonians. In many chelonian populations which typically reproduce annually, some adult females still may not lay eggs in a given year (e.g. *Deirochelys reticularia*: Gibbons and Green 1978; *Gopherus polyphemus*: Auffenberg and Iverson 1979; Landers et al. 1980; *Chrysemys picta*: Tinkle et al. 1981; *Emydoidea blandingii*: Congdon et al. 1983; *Chelydra serpentina*: Congdon et al. 1987; *Trachemys scripta elegans*: Frazer et al. 1990; *Pseudemydura umbrina*: Kuchling and Bradshaw 1993). However, the fact that some females skip reproduction during a given breeding season does not necessarily demonstrate that multi-year reproductive cycles are inherent to a population, it may rather indicate the ability of females to switch off reproduction under particular environmental or nutritional constraints. Females may not lay eggs in a given year either because they do not initiate or sustain a vitellogenetic cycle and lack follicles of preovulatory size (as is the case in sea turtles which have multi-year ovarian cycles), or because they fail to ovulate despite a normal vitellogenetic cycle. The second scenario, in which females show annual cycles of vitellogenesis and follicular enlargement, but abort ovulation and egg production in a given season, occurs occasionally in the Western Australian freshwater turtles *Pseudemydura umbrina* (Kuchling and Bradshaw 1993) and *Chelodina steindachneri* (G. Kuchling and S.D. Bradshaw, in prep.). It has to be stressed that these females do show normal, annual ovarian cycles, even though they may not lay eggs on an annual basis. Most population studies of non-marine chelonians during which, in some years, individual females could not be observed to be gravid or nesting (Gibbons and Green 1978; Auffenberg and Iverson 1979; Landers et al. 1980; Tinkle et al. 1981; Congdon et al. 1983; Congdon et al. 1987; Frazer et al. 1990) did not investigate the ovarian cycles (for example, by endoscopy or ultra-sound scanning) and, therefore, do not provide any information on the underlying patterns of gonadal activity.

According to the hypothesis that the driving forces for multi-year reproductive cycles in marine turtles are the energetically demanding breeding migrations in combination with constraints of energy acquisition, the prime candidates for multi-year gonadal cycles in non-marine chelonians would be some of the large river turtles which produce relatively large numbers of eggs, and which show reproductive migrations comparable to those of the sea turtles (see Sect. 4.1). Recently, Junk and Silva (1997) presented a table containing the information that, in the central Amazon flood plain, females of the large (up to 80-cm shell length), migratory arrau turtle *Podocnemis expansa* lay 48–134 eggs every 2–4 years. However, Junk and Silva (1997) neither presented supportive data nor references for their claim, nor did they mention it in their discussion of the reproductive biology of *P. expansa*. They also did not specify whether one or several clutches are laid during an active breeding season. Some years earlier, after a comprehensive review of reproductive data, Pritchard and Trebbau (1984) had concluded that females of *P. expansa* lay one clutch on an annual basis.

A general problem in non-marine chelonians seems to be the correct determination of multi-year gonadal and reproductive cycles. Even if conclusions are based on an assessment of the condition of ovaries, there seems to be uncertainties and confusions about how to interpret ovarian conditions in regard to multi-year cycles. For example, Pritchard and Trebbau (1984) based their hypothesis, that individual females of *P. expansa* nest once per year on an annual basis, on the premise that enlarged follicles observed in the ovaries of post-nesting females represent clutches to be shelled and deposited in subsequent years. Although persistence of vitellogenic follicles until the following year's breeding season does occur in some temperate-zone turtles (e.g. *Clemmys guttata*: Ernst and Zug 1994), such a pattern has not yet been demonstrated in any tropical species. Without presenting any data or details about how they came to their conclusion, Thorbjarnarson et al. (1993: 345) stated that "the examination of ovaries of females killed during the dry season in the Capanaparo indicates that both *P. unifilis* and *P. expansa* do not nest more than once annually". Unfortunately, they did not explain the basis for their reasoning.

Erymnochelys madagascariensis, a relatively large (up to 0.5 m shell length) river turtle from western Madagascar belongs to the same family Podocnemidae as the South American river turtles of the genera *Podocnemis* and *Peltocephalus*. In the rare and endangered *Erymnochelys*, the analysis of ovarian data of twelve adult females revealed a biennial ovarian cycle according to the following criteria: (1) only 50% of the adult females showed ovulation scars (corpora albicantia) on the ovaries after the peak nesting season (Fig. 5.1A) and half a year later (Fig. 5.1B); (2) the other adult females lacked ovulation scars and preovulatory follicles during the peak nesting season, but had batches of small follicles which appeared to be recruited for the following year (Fig. 5.1C); (3) half a year

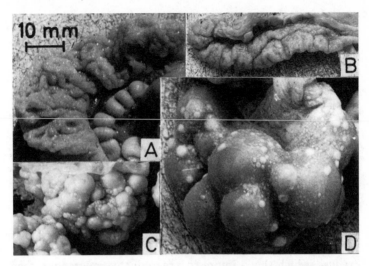

Fig. 5.1A–D. *Erymnochelys madagascariensis.* Formalin-preserved ovaries at different stages of the biennial reproductive cycle. **A** November, post-reproductive ovary, corpora albicantia of 3–6 mm and of about 2 mm. **B** May, ovary regressed, old ovulation scars of 1–1.5 mm. **C** November, early vitellogenesis, follicle size classes with diameters of 5 and 3–4 mm. **D** May, late vitellogenesis, follicle size classes with diameters of 17–18, 15, 12–13, 5–7 and 3–4 mm

after the peak nesting season, enlarged vitellogenic follicles, the prospective eggs for the next reproductive season, were only found in females without ovulation scars from the previous nesting season (Fig. 5.1D). Females lay multiple clutches from September to January, at the end of the dry season and the beginning of the wet season. Ovaries regress dramatically after the breeding season and no vitellogenesis occurs until the following October/November when follicles slowly start to enlarge. Vitellogenesis then progresses over a whole year until the next breeding season (Fig. 5.2). Females, therefore, lay several clutches of eggs during an active breeding season, which, presumably, occurs every second year (Kuchling 1993a).

Erymnochelys is omnivorous, with females in the flood lakes of large rivers being mainly herbivorous during the wet season and carnivorous during the dry season when their food consists mainly of aquatic snails (Kuchling 1993c). No breeding migrations are known for *Erymnochelys*. Although the biennial cycle of *Erymnochelys* may be related to the energetic constraints of their alimentary regime, this in itself does not explain why a similar energy investment per egg should not be obtained by producing less or smaller clutches every year instead of two or three large clutches every second year. Reduced annual egg numbers have, for example, been observed in populations of *Emydura krefftii* (Chelidae) inhabiting unproductive environments (Georges 1985). However, Bull and Shine (1979) speculated that yearly reproduction, although advantageous in good habitats, may become less favourable as the habitat becomes poorer. There may also be a phylogenetic component in the biennial ovarian pattern of *Erymnochelys*, e.g. its ancestors may have shown reproductive migrations. The biennial cycle in

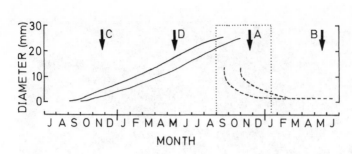

Fig. 5.2. Biennial ovarian cycle of *Erymnochelys madagascariensis*. *Solid lines* Aassumed diameter of the two largest size classes of follicles; *dashed lines* size of ovulation scars; *stippled area* time of ovulation and oviposition; *arrows* and *letters* stages shown in Fig. 5.1

Erymnochelys suggests that the hypothesis of a multi-year breeding cycle (Junk and Silva 1997) is more plausible for the related, South American *Podocnemis expansa*, which shows impressive reproductive migrations, than Pritchard and Trebbau's (1984) alternative assumption that the females lay eggs on an annual basis. Due to the variety of life styles and ecology in the family Podocnemidae, the six species of *Podocnemis* and the monotypic genera *Peltocephalus* and *Erymnochelys* potentially offer excellent models to investigate the relationships among migration, colonial or solitary nesting, omnivory or herbivory and frequency of reproduction. Junk and Silva (1997) reported that *Podocnemis unifilis, P. erythrocephala* and *P. sextuberculata* all lay eggs once per year, and *P. vogli* reputedly produces three annual clutches (Ramo 1982, cited in Thorbjarnarson et al. 1993), but I consider the reproductive data for this group still to be insufficient to evaluate the patterns of ovarian activity.

The reproductive cycles of the large, threatened, migratory Asian river turtles of the family Bataguridae are even less known than those of the Podocnemidae. Since the possibility of multi-year cycles has implications for their reproductive output and fecundity (and, therefore, for conservation management), this aspect should receive more attention in future investigations.

5.4 Unpredictable Habitats

The discussion of the various reproductive cycles above concerns habitats in which the timing of the environmental fluctuations is regular and predictable, although year to year variations in onset, magnitude and duration of, for example, rainy seasons or cold winter temperatures certainly occur. Overall, however, there is a general, predictable pattern that it will be colder in winter or wetter during the rainy season. How do chelonians reproduce when the occurrence of essential environmental parameters, e.g. rainfall, becomes unpredictable and, with it, the timing of the "ultimate causes" for particular breeding seasons *sensu* Baker (1938) – those periods of the year that are most propitious for the survival of both parent and young? In extreme habitats, for example desert habitats, resources like food and water may be very patchily distributed in space and time, particularly for aquatic turtles. Temperature generally shows seasonal fluctuations, but, depending on the type of desert, rainfall may or may not follow a seasonal pattern. Some (and there are only a few) freshwater turtles living in temporary waters in desert areas may be confronted with a high degree of unpredictability for their essential aquatic habitat.

A large part of Western Australia is arid (mean annual rainfall 200–250 mm), with only the south-western and northern fringes receiving higher rainfall. Although most Western Australian freshwater turtle species (family Chelidae) live in the higher rainfall areas of the south-west (mediterranean climate) or the north (tropical summer monsoon climate), the long-necked turtle *Chelodina steindachneri* is only found in the arid country. *Chelodina oblonga* lives in permanent or semi-permanent aquatic habitats in south-western Australia. *Pseudemydura umbrina* also occurs in south-western Australia, but in ephemeral clay swamps. A comparison of the reproductive strategies of these three species provides insight into how chelonians can (or cannot) adjust their reproductive cycles to environmental unpredictability.

The basic gonadal cycles of all three species are very similar: vitellogenesis starts during summer (February), and the typical nesting seasons are spring to early summer (Kuchling 1988b; Kuchling and Bradshaw 1993). All three species have the possibility to either initiate a vitellogenetic cycle, which all females do under normal circumstances (all three species typically show annual gonadal cycles), or, if external or internal conditions are adverse, not to initiate a vitellogenetic cycle. *Chelodina oblonga* represents the conventional freshwater turtle reproductive pattern, whereby, once preovulatory follicles are present at the start of the breeding season, ovulation occurs and eggs are laid at the normal time of the year. This also happens in unusually dry years when particular females have not been able to feed for many (e.g. 10) months or in experiments simulating dry conditions in spring. Once *C. oblonga* females have initiated a reproductive cycle, which happens 8–10 months prior to the breeding season when vitellogenesis starts, their decision to breed has been made. Although it is normal in this and many other species that, during and after the breeding season, some follicles become atretic and are reabsorbed (see Sect. 3.2.6), *C. oblonga* does not use this mechanism to abort reproduction if environmental conditions are suboptimal during the breeding season.

Pseudemydura umbrina inhabits temporary swamps, the water levels of which depend largely on rainfall. Rainfall in the area of its distribution is concentrated in late autumn and winter and, with some predicability, standing water is available in *P. umbrina* habitat during winter and early spring, but not in other seasons. The unpredictability of the habitat mainly concerns the amount of rainfall and the duration of aquatic swamp life, e.g. there are good and bad years. Regarding the reproductive cycle, it is important to note that vitellogenesis starts during aestivation in summer, at least half a year before water becomes available and 8 or 9 months before it becomes obvious if it is a bad or a good season for breeding. *Pseudemydura umbrina* feeds underwater on live aquatic prey, then aestivates (and does not feed) from late spring to late autumn or early winter (typically 6–8 months) in underground holes or under leaf litter. In occasionally occurring low-rainfall years, when water may be available only for a few weeks and the aestivation period is correspondingly extended, females do not ovulate and, by late spring, all ovarian follicles become atretic and are reabsorbed (Kuchling and Bradshaw 1993). The actual trigger for ovulation seems to be a high food intake in early spring (see Sect. 6.1.3). This feeding peak can only occur when water is available and swamp life is plentiful during early spring.

Not only does *P. umbrina* have the ability to abort ovulation and to reabsorb the yolk and its stored energy if the environment is suboptimal, but also gravid females may, if conditions are stressful, abort soft-shelled eggs which are not yet ready for oviposition. They simply drop them in the water or on the ground (Fig. 5.3). At first glance, abortion of soft-shelled eggs does not seem particularly adaptive, since energy reserves are wasted which are normally invested in reproduction or which are reabsorbed and available to the female during aestivation. However, gravidity also places other than energetic constraints on *P. umbrina* females: due to the congestions caused by the oviducal eggs in the body cavity, gravid females cannot fill their bladders and accessory cloacal bladders with water (which can be easily assessed by ultra-sound scanning). If no water is available to the females after nesting and oviposition, they have to start aestivation with low water stores, a situation which is potentially lethal during drought years when the aestivation period is particularly long. In evolutionary terms, the increased survival chances of a gravid female which aborts her eggs under drought conditions and which re-fills her water stores may counterbalance the loss and waste of the energy of some eggs.

These mechanisms of *P. umbrina* to cancel reproduction in the short term seem adaptive to its ephemeral habitat after taking into consideration the fact that vitellogenesis starts long before environmental adversities during the breeding season become obvious. Females have several lines of decision making to follow on whether to breed or not to breed in a particular year (Table 5.2).

Some of the populations of *C. steindachneri* inhabit ephemeral river pools in subtropical, arid areas which receive most of their precipitation and water runoff from cyclonic rains. *Chelodina steindachneri* feeds and mates only in water and aestivates when no water is available (Burbidge 1967). In more extreme habitats, for example Borodale Creek at the edge of the Great Victoria Desert, water may only be available after heavy rains for periods of 4–6 weeks at a time. Rains in this area (annual mean about 200 mm, with a range from 35 mm to 690 mm between

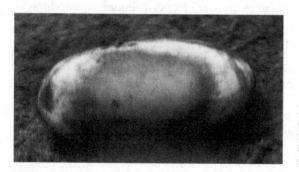

Fig. 5.3. Aborted soft-shelled egg of *Pseudemydura umbrina*, found underwater; the yolk is shining through the shell membrane

1940 and 1965) may occur at any time of the year, with the likelihood of rain being greatest in summer and early winter, and with the probability of drought in the spring months of September and October. Rains, generally, are infrequent and their timing is unpredictable. Rainfall, too, has to reach a certain intensity to provide standing water in the creek beds and pools. Although no data are available on water levels of creeks, calculations of "effective rain" events for plant growth, which is the precipitation adequate to initiate plant growth in an area (which also depends on the temperature and, therefore, varies from 6 mm in July to 28 mm in January), give some indication on the availability of water. Between 1940 and 1960, intervals between effective rain events in the wider area of Borodale Creek varied from 3 to 376 days and averaged three events per year. It exceeded 300 days four times and 200 days eight times. Longer (>1 year) droughts also occur intermittently. During the 40 months from January 1961 to April 1964, only one effective fall of rain was received. Maximum air temperatures during the summer half-year (November to March) are regularly in the high 40s (°C) and minimum temperatures during winter (May to September) below zero (Beard 1974).

Borodale Creek may be one of the most unpredictable and extreme habitats for any freshwater turtle. Preliminary results of a reproductive biological study of three different *C. steindachneri* populations, including the one at Borodale Creek, reveal that *C. steindachneri* females nest in late September or early October if water is available. When water remains available, a few females may even lay a second clutch in November. If water is not available in spring during their normal breeding season, *C. steindachneri* females, like *P. umbrina* females, do not ovulate. However, contrary to *P. umbrina* in which follicles that do not ovulate become atretic during late spring, *C. steindachneri* females maintain preovulatory follicles and follicles in the late vitellogenic state in perfect condition, ready for ovulation, until at least January (Kuchling 1988b; G. Kuchling and S.D. Bradshaw, in prep.). Even if their habitat is dry during spring, rain may well fall and water and food may become available during summer, a situation which is highly unlikely in the habitat of *P. umbrina*. *Chelodina steindachneri* females may

Table 5.2. To breed or not to breed or when to breed: stages of decision making in southwestern Australian freshwater turtles

	C. oblonga	P. umbrina	C. steindachneri
Initiate or do not initiate vitellogenesis	+	+	+
Ovulate or do not ovulate preovulatory follicles during breeding season	–	+	+
If stressful, abort soft-shelled eggs	–	+	–
If pre-ovulation, retain late vitellogenic follicles beyond normal breeding season	–	–	+
If necessary, retain eggs in oviduct until conditions for nesting are favourable	–	–	+

+ Decisions about alternatives possible; – not possible.

store enlarged follicles and delay egg production for many months if environmental conditions are unfavourable.

We (G. Kuchling and S.D. Bradshaw, in prep.) did not find any indication that, under stressful conditions, gravid *C. steindachneri* abort oviducal eggs. If nesting conditions are lacking, for example because females are kept in a water tank without access to land, gravid *P. umbrina* and *C. oblonga* females with fully shelled eggs will, after their species-specific oviducal period plus a margin of a few days is completed, simply drop their eggs in the water. Wild *P. umbrina* females sometimes start aestivation while in a late gravid state and have to interrupt aestivation, construct a nest and oviposit when their normal oviducal period of 39–49 days is completed. This seems maladaptive, in particular since the dry environment does not allow the replenishment of the turtle's water stores (see above). In contrast, gravid *C. steindachneri* females may, if nesting conditions are lacking, keep their shelled eggs for at least 4 months (until February, but potentially longer) in the oviducts and only nest when suitable nesting conditions become available. During that time, the properties of the egg shells, as assessed by ultra-sound scanning, remain unchanged. By holding back eggs and storing them in the oviducts without changes to their shell, gravid *C. steindachneri* may avoid the necessity to interrupt aestivation in order to nest and oviposit under stressful environmental conditions, an option which *P. umbrina* does not have (and a reason why *P. umbrina* may occasionally use the option to abort soft-shelled eggs).

It is interesting to note that the spermatogenetic cycle of *C. steindachneri* is more or less similar in its timing to that of *C. oblonga* (Kuchling 1988b). Spermatogenesis proceeds and peaks during summer no matter if water is available and the turtles are active or if it is dry and turtles have to aestivate. The capacity to store sperm allows males to reproduce in unpredictable environments without any need to modify mechanisms.

To summarise adaptations of south-western Australian turtles to unpredictable habitats (Table 5.2), *P. umbrina*, which inhabits seasonal, ephemeral swamps, has the ability to abort reproduction in the short term if environmental conditions are unsuitable and stressful. *Chelodina steindachneri*, some habitats of which could be described as erratic, has the ability to delay ovulation and egg production for many months until conditions become suitable and, thus, shows a remarkable flexibility in the timing of reproduction. In addition, gravid females may also delay oviposition for many months and store eggs in the oviducts to wait for suitable nesting conditions. These mechanisms allow *C. steindachneri* to thrive and reproduce in extreme, unpredictable habitats without a need to modify its basic gonadal cycle which is very similar to that of related species. The novelty lies in a shift of a rather narrowly timed pattern of egg production, which seems to be typical for most chelonians, to a labile or open pattern which allows these species to adjust the timing of egg production and oviposition to environmental vagaries.

Similar mechanisms may also occur in some other chelonians experiencing unpredictable environments, but few cases are documented. One main reason for the lack of knowledge about these mechanisms in species other than those discussed above is the fact that few researchers use techniques which allow the assessment of ovarian conditions in live chelonians, such as ultra-sound scanning or endoscopy. Many field researchers restrict their non-lethal techniques to radiography which limits the assessment to determining the occurrence of shelled oviducal eggs. Retention of fully shelled oviducal eggs is the only one of the mechanisms discussed above which is also known to occur in other species (see Sect. 3.2.5). Retention of oviducal eggs over many months seems to be common in North American chicken turtles *Deirochelys reticularia* which inhabit ephemeral, fluctuating aquatic habitats (Buhlmann et al. 1995; Buhlmann 1997). In this species, nesting occurs either in autumn or winter or eggs are retained in the oviducts until spring. The reason for egg retention in *D. reticularia* may be to wait for suitable nesting conditions.

A radiographic study of reproduction of a population of the western or desert box turtle *Terrapene ornata luteola* in a harsh and unpredictable environment in New Mexico offers indirect indications that some mechanisms of egg production may operate similarly to those of Australian chelids inhabiting unpredictable environments. These data indicate that spring rains may increase the proportion of *T. o. luteola* females laying eggs during summer – and that in dry years, turtles may defer laying eggs completely (rather than reducing annual output). Retention of shelled eggs in the oviducts can occur for at least 50 days (Nieuwolt-Dacanay 1997). These are hints that *T. o. luteola* may also have the ability to abort ovulation and egg production during dry conditions and to delay oviposition. In contrast, the terrestrial desert tortoise *Gopherus agassizii* may have different strategies to cope with environmental unpredictability. Desert tortoise females tend to reproduce every year despite variability in rainfall and food availability, even in drought years. Their strategy is to forfeit body condition to produce only a few eggs, even under great environmental stress (Henen 1997), although, in years following drought, post-nesting body condition is to some degree correlated with future reproductive output (Karl 1997). It would be interesting not to limit those investigations of the American desert-dwelling *G. agassizii* and *T. o. luteola* to counting shelled eggs by radiography, but to study by ultra-sound scanning or endoscopy the underlying gonadal mechanisms which determine egg production under different environmental conditions.

5.5 Conclusions

Although climatic zones and environmental conditions do exert influences on gonadal cycles of chelonians, the basic patterns of gonadal activity change little and are surprisingly uniform. Vitellogenesis, the energy allocation for the prospective eggs, typically takes many (8–12) months in all the species which were investigated, and may be extended over years in some of the northernmost species. Spermatogenesis seems to peak during the warmest period of the year regardless of the timing of activity, mating and nesting. In chelonians, the extensive potential for sperm storage in males and females (see Sects. 3.3.6 and 3.5) does not necessitate the synchronisation of male and female gonadal cycles. It enables the temporal dissociation of mating and insemination not only from spermatogenesis, but also from ovulation and fertilisation. Chelonian reproductive cycles, thus, have traits which can be regarded as "exaptations" (*sensu* Gould and Vrba 1982: a trait that improves the fitness of an organism without having evolved in situ through the action of natural selection) to function in various climatic zones, and may not need any (or only minimal) further adaptations to particular environmental conditions.

Ovulation and oviposition are the main events which determine the seasonal environmental conditions for egg incubation. Ovulation and oviposition also define the timing of the restrictions which gravidity and nesting places on females. Adaptations (which arose in a given environment through the action of natural selection for the particular trait, i.e. by differential survival of organisms with high expected fitness: Gould and Vrba 1982; Bradshaw 1997) of reproductive mechanisms to unpredictable fluctuating environments include the sensitivity of ovulation to particular external or internal triggers and, in a few chelonians, the dissociation of the temporal relationship between ovulation and oviposition through variation of the duration of gravidity.

In some species, the relationship of oviposition to the optimal time for hatchling emergence is also rather loose, because additional factors regulate the timing of this event: embryonic development may be temporarily arrested and/or hatchlings may go into dormancy before hatching and/or before emerging from the nest (see Sects. 4.4 and 8.2). Chelonians, therefore, exhibit a remarkable flexibility in meeting the ultimate cause for the timing of reproduction (*sensu* Baker 1938) under various climates without strong selective pressure to shift or restructure basic gonadal cycles and their proximate cues. This flexibility and adaptability of their basic reproductive cycles, to function under various climatic and environmental conditions, may have contributed to the success of chelonians with regard to persistence through geological times and changing climates.

Chapter 6

Control of Reproduction

The first chapters of this book have shown that reproduction is an extremely complex phenomenon which includes maturation of the gonads, development of primary and secondary sex characteristics, gamete production, preparation of the reproductive ducts, and reproductive behaviour. The previous chapter described the various reproductive cycles in different climatic zones and habitats and demonstrated that, for most if not all chelonians, reproduction is a seasonal and cyclic event. The focus of this chapter is on the "proximate causes" of breeding seasons *sensu* Baker (1938), on the factors that initiate the sexual cycles at the appropriate time of the year and on the mechanisms through which these factors regulate the processes of reproduction.

In most organisms, changes in reproductive activity can be induced by external environmental conditions which act as proximate cues. Until the middle of the twentieth century, the general view was that the annual reproductive cycles were directly driven by seasonal cycles of environmental factors. Aschoff (1955) developed a conceptual framework for annual rhythms of reproduction in homeotherm vertebrates which considers environmental factors as *zeitgeber* that synchronise endogenous circannual rhythms, rather than as factors directly causing seasonal biological cycles. A large body of research in mammals and birds demonstrates that many particularities of both photoperiodic actions and spontaneous seasonal changes can indeed be better understood on the basis of this general model (Gwinner 1986). Whittier and Crews (1987) view the mechanisms that control seasonal reproduction as lying on a continuum between two extremes: pre-programmed or closed control (endogenous cycles) versus labile or open control (exogenous cues). In reptiles, the temporal control mechanisms for reproduction combine exogenous and endogenous regulators and numerous intermediate conditions occur (Duvall et al. 1982). In lizards, for example, one part of the annual cycle may be endogenously timed while another phase may depend on exogenous cues (Fischer 1974; Licht 1984). Physiological studies relating to timing mechanisms in chelonians are still scarce.

6.1 Exogenous Factors

A variety of environmental factors, such as temperature, photoperiod, rainfall, moisture, humidity and food supply, as well as sexual behaviour and social interactions may act as proximate cues to regulate reproduction in reptiles (Duvall et al. 1982). Environmental factors may directly trigger physiological processes, they may be permissive factors for physiological processes and for the expression of endogenous circannual rhythms, or they may be external *zeitgeber* to synchronise circannual rhythms with the time of the year. Although most field studies discussed in Chapter 5 attempted to correlate gonadal changes with climatic variables, these descriptive studies provide little insight into the physiological mechanisms which synchronise chelonian reproductive cycles. Few experimental studies have tested which environmental parameters are involved in controlling chelonian reproduction.

6.1.1 Temperature and Photoperiod

Temperature and photoperiod are conspicuous and regular features of climatic seasonality in temperate regions. The annual change in daylength, in particular, has a unique regularity compared to temperature which may show pronounced short-term variations and fluctuations. Photoperiod evolved as the primary proximate factor regulating reproduction in birds (Follett 1984) and many other organisms. Temperature, however, is of direct relevance for the ectotherm reptiles, with cold winter temperatures slowing down their physiological processes, and higher temperatures (i.e. the thermal preferendum of the species) accelerating them. In lizards, the best studied group of reptiles concerning photoperiod and temperature effects on reproduction, temperature offers distinct timing cues for various phases of gonadal cycles in most species, but evidence that daylength acts as an important proximate cue for gonadal cycles is meagre and restricted to a few species (Licht 1984).

Male and female gonadal cycles of chelonians are generally not synchronised (see Chapter 5). In all temperate zone species, spermatogenesis starts in spring and spermiogenesis and spermiation reach a peak in summer or early autumn before a period of germinal quiescence occurs from late autumn to early spring. Temperate zone females, on the other hand, typically show a more or less pronounced period of ovarian quiescence during summer, with vitellogenesis starting in late summer or autumn and progressing (with a break during hibernation) until spring, when ovulation and oviposition occur (see Sect. 5.1). It can, therefore, reasonably be expected that photoperiod and temperature affect male and female gonads of chelonians in different ways.

Burger (1937) investigated experimentally from November to February the role of photoperiod and temperature on spermatogenesis of *Trachemys scripta elegans* (Emydidae), and concluded that photoperiod was the most important factor for inducing a new spermatogenetic cycle. However, these aquatic turtles were kept under suboptimal conditions, in terraria with only limited access to shallow water which did not even cover their shells, and the animals were accordingly in poor health. Artificial lighting only was provided to the experimental group – and by 50-watt bulbs which, since the turtles were basically kept terrestrially, may also have acted as a source of radiation heat. Burger's speculation that increased illumination and not temperature caused the induction of a new spermatogenetic cycle is rather unconvincing and, at best, ambiguous.

In the terrestrial tortoise *Testudo hermanni*, a drop in the environmental temperature was found to be the main factor inducing testicular regression during autumn. Elevated temperatures during winter advanced the beginning of a new spermatogenetic cycle, but photoperiod had no effect during the early phases of the spermatogenetic cycle (spermatocytogenesis). Suboptimal temperature during summer – a constant temperature of 21 °C – inhibited spermiogenesis, whereas spermiogenesis proceeded at a constant temperature of 26 °C. In addition, a long photoperiod possibly had a stimulatory effect on spermiogenesis. At 26 °C and with a light:dark cycle of 16:8, spermiogenesis proceeded strongly until the end of November and testes did not regress (Kuchling 1982a).

Ganzhorn and Licht (1983) found that, in males of *Chrysemys picta* (Emydidae), temperatures at or below 17 °C completely suppressed the testicular cycle. Full spermiogenesis was induced at 28 °C within 7 weeks in either autumn or spring. However, in contrast to the results in *T. hermanni* (Kuchling 1982a), males of *C. picta* exhibited spontaneous testicular regression even under constant warm temperatures. Daylength had no apparent effect on testicular activity of *C. picta*, neither in autumn nor in spring.

In the North American musk turtle *Kinosternon odoratum* (Kinosternidae), temperature and photoperiod influence the male cycle in a similar way to *T. hermanni*. Mendonça and Licht (1986b) found elevated temperature to be the major environmental factor affecting testicular recrudescence, spermatocyto-genesis, spermiogenesis and androgen secretion in *K. odoratum*. Photoperiod, although interacting with temperature, influenced only spermiogenesis and testicular regression in autumn.

The effects of temperature and photoperiod on ovarian activity have been little explored experimentally. Ganzhorn and Licht (1983) found that, in *Chrysemys picta*, warm temperatures (28 °C) inhibited follicular growth and caused regression of ovaries. Maximal ovarian growth and ovulation were observed only in spring in females kept at a constant water temperature of 17 °C, with or without several hours of basking daily. Follicles also grew at lower temperatures (13 °C), but ovulation did not occur. Although Ganzhorn and Licht (1983) did not demonstrate any influence of daylength, they speculated that increasing or long days might be important for the final stages of the ovarian cycle (ovulation) in spring. However, no apparent effect of photoperiod was detected in any season on the ovarian cycle of *Kinosternon odoratum* (Mendonça 1987a). In these experiments, high temperature (28 °C) was initially stimulatory in autumn when it increased ovarian mass due to an increase in the number of preovulatory follicles, but after 10 weeks of 28 °C, follicles, instead of ovulating, became atretic. If females were continuously exposed to high temperature for 7 months, they exhibited an increase in ovarian mass, but this gain was due to a higher number of smaller follicles rather than preovulatory ones, indicating abnormal growth. Generally, prolonged exposure to high temperature inhibited follicular growth also in *K. odoratum*, whereas a lower temperature (18 °C) promoted it. Ovulations were rare, and only occurred at the low temperature treatments.

Earlier observations of egg production in a captive group of southern European tortoises of the species *Testudo marginata*, *T. graeca ibera* and *T.*

hermanni in Germany between 1951 and 1961 indicate that environmental temperature may be an overriding factor in regulating female reproductive cycles, although the environmental conditions in this experiment were not closely controlled or monitored. From 1951 to May 1960 this group of tortoises was kept in a greenhouse with infrared lamps running for 10–12 h daily and, although under warmer conditions than are natural in Germany, were thus still exposed to the seasonal and daily temperature fluctuations. Eggs were obviously laid during late spring and early summer. In May 1960 the group was transferred to an indoor facility (and still kept under infrared lamps) and eggs were laid in early summer of 1960. During the winter of 1960/1961, room heating increased the basic room temperature to about 25 °C and the tortoises produced a total of 62 eggs between 8 January and 6 April 1961, with some females laying three clutches during this period. No eggs were laid in spring after 6 April and over the summer of 1961. The room heating was turned on again in September 1961, increasing the basic room temperature from 22–24 °C to about 25 °C. This temperature increase was followed by a total of 104 eggs being laid between 24 September and 16 December 1961, again with some females laying three clutches (Rohr 1970). Elevated temperature, thus, can advance the egg laying period in southern European tortoises by several months, and a modified annual temperature cycle can shift the phase and speed up the female reproductive cycle.

The only tropical turtle in which the effects of photoperiod and temperature on ovarian activity have been experimentally tested is the Indian softshell turtle *Lissemys punctata*. Sarker et al. (1996a) tested temperatures which were 3 °C lower than the lowest ambient temperature and 3 °C higher than the mean highest ambient temperature of the respective phase. They found no effect of photoperiod during any phase of the ovarian cycle. Short-term (4 weeks) high temperature treatment significantly stimulated ovarian growth and secretion of oestradiol-17β during the preparatory phase (March), whereas after 10 weeks at that treatment follicles became atretic. During vitellogenesis (recrudescent phase, July) high temperature caused degeneration of follicles in the short or long term. Low-temperature treatment slowed down follicular growth and oestradiol-17β during the preparatory phase as well as during vitellogenesis. The quiescent ovary (December) was stimulated by neither high nor low temperatures. Thus, elevated temperature seemed to trigger follicular growth and oestrogen secretion in the early preparatory phase, at the same time as when temperatures also rise in the wild habitat, but during the phases of vitellogenesis and ovulation all treatments seemed to cause degeneration of follicles. Since no control groups in the same experimental tanks under the normal seasonal temperatures were provided, it cannot be evaluated whether the temperature treatments or general stress factors (see Sect. 2.5) were responsible for this outcome.

To summarise, the above studies demonstrate that temperature represents the major proximate cue for gonadal activity in chelonians, with photoperiod having only a limited additional role in some phases of the gonadal cycles. However, low temperatures that suppress the testis cycle of chelonians may stimulate vitellogenesis and, vice versa, high temperatures which are necessary for spermiogenesis may cause regression of ovaries. Ovulation and oviposition may depend on increasing temperatures, but more so in the middle than in the high range. This relationship reflects well the dissociation of the male and female gonadal cycles of temperate zone chelonians.

Experimental data on the influence of temperature and photoperiod on chelonian reproductive behaviour are even scarcer than those on gonadal cycles. Mendonça (1987b) demonstrated experimentally that short daylength was the key environmental condition stimulating male sexual behaviour in *Kinosternon odoratum*, but temperature fluctuations in the microhabitat (weather changes) were also stimulatory and temperature changes fine-tuned the behaviour. Therefore, although photoperiod may only play a rather limited role in the control of gonadal cycles of chelonians, it may still be an environmental cue for their reproductive behaviour.

6.1.2 Water Levels, Rain and Moisture

There is good evidence that many tropical and subtropical chelonians time reproduction to coincide with either the dry or the wet season (see Sect. 5.2). Explanations of ultimate causes for dry season nesting include the availability of drier, warmer nest sites and improved egg survival, and the risk of inundation of nesting beaches during the wet season (Moll 1979). Less clear are the proximate factors which regulate the timing of reproductive cycles with rain, moisture or water levels of rivers. Experimental data are still lacking or insufficient.

In *Podocnemis expansa*, in which nesting takes place when the rivers have their lowest water levels and sand banks are exposed, females enter a phase of basking on sand banks before the actual onset of nesting. This may be the only time during their life cycle that individuals of *P. expansa* engage in terrestrial basking (Pritchard and Trebbau 1984). Basking in *P. expansa* may play an important role in maturation and ovulation of ovarian follicles and/or in the shelling of oviducal eggs and, therefore, in the timing of oviposition which differs in different rivers or tributaries (see Sect. 5.2). Basking behaviour can only start when river levels have fallen enough to expose sand banks and, thus, offers a reliable timing cue for low river levels and when to produce eggs. The actual trigger may either be the increased body temperature or the increased light intensity during basking or both, but experimental data are lacking.

Stettner (1996) observed that he could regulate the mating activity of a captive *Chelodina mccordi* male, a tropical species, by offering either wet or dry season conditions: the male exhibited strong mating activity under relatively low illumination, moderate water temperature (25 °C) and saturated air humidity (wet season condition), but stopped mating attempts under a regime of intense illumination of 12 h daily, a water temperature of 28–30 °C and lower air humidity (dry season condition). The combined change of several parameters, however, precluded the evaluation of which factors were ultimately responsible.

It has to be kept in mind that chelonian gonadal cycles take between half a year and three quarters of a year, and that rain or moisture or the lack of it may trigger the last phases of the reproductive cycle rather than initiate a new cycle. It may well be that many aquatic, tropical species use temperature and light intensity as indicators for wet-season or dry-season conditions rather than rain or moisture per se. In *Lissemys punctata*, the only tropical species in which the effects of temperature and photoperiod where experimentally tested (see previous Sect.), temperature seemed to be the most important proximate cue (Sarker et al. 1996a). In future experiments of this kind with tropical species, it may be worthwhile to control for illumination intensity as well as – or instead of – for photoperiod. There is, however, also the possibility that rain, moisture, light and temperature are not the main triggers, but that other factors which are related to them act as proximate cues, for example food availability.

6.1.3 Food Availability

Food availability, which often depends on rainfall and moisture, may also act as a timing cue for reproductive events in chelonians. A correlation between body condition, which depends to a large degree on the amount of food available, and reproductive output (egg numbers) occurs when food or energy acquisition is limited. This is particularly obvious in herbivorous species. In the herbivorous *Geochelone gigantea* on Aldabra, reproductive output (numbers of ovarian follicles and/or eggs per female) is strongly influenced by primary production, which is in turn influenced by rainfall (Swingland and Coe 1978; Swingland and Lessells 1979). In higher rainfall years, females lay more eggs. However, the question of interest here is if food availability per se can switch reproduction on or off, rather than only modulate it. This may at least partly be the case in green turtles *Chelonia mydas*, in which the lag period between the multi-year ovarian cycles varies according to the growth of seagrass and algae (see Sect. 5.3). Experimental evidence, however, is lacking. A likely explanation is that the initiation of vitellogenesis, rather than being directly triggered by food availability, depends on reaching a certain body condition threshold. The trigger for the "decision" to breed or not to breed in a given year may be a certain level of fat reserves (Miller 1997), which can be regarded as an endogenous, permissive factor. In species with multi-year ovarian cycles, females may not be able to reach the body condition threshold to start vitellogenesis in the year following an active breeding season. Their body condition simply may not permit the start of a new cycle.

To face the energy constraints of reproduction, the giant tortoise and the green turtle have obviously developed contrasting strategies. The migratory green turtle, compared to the giant tortoise, has larger accessory energy expenditures of reproduction (through its long migrations) which are relatively independent of fecundity per active breeding season (see Sect. 5.3). If chelonian groups were ranked according to "capital" and "income" breeders, as has been proposed for birds (Drent and Daan 1980), the green turtles would certainly be at the far end of the capital breeders, which only initiate an ovarian cycle if its capital, in the form of energy reserves, is high enough to realise its full potential of fecundity. The giant tortoise could be classified as an income breeder, which may have a relatively lower condition threshold to initiate breeding and which adjusts its energetic expenditure to low energy income by reducing the egg numbers per breeding season. However, a precise body condition threshold for breeding, as it has been demonstrated and defined for the viviparous snake *Vipera aspis* (Naulleau and Bonnet 1996), has not been determined for any chelonian.

In *Pseudemydura umbrina*, ovulation only occurs after a feeding peak in early spring. Circumstantial evidence suggests that the lack of an elevated food intake in spring was the main reason why captive females of this critically endangered species did not produce any eggs from 1982 to 1986. The females, then, showed normal annual cycles of vitellogenesis, but did not ovulate and reabsorbed the follicles (Kuchling and DeJose 1989; Kuchling and Bradshaw 1993). It has to be stressed that these captive females were in a generally good condition, clearly over the condition threshold to initiate vitellogenesis. However, a good, static level of fat reserves or general body condition are not sufficient to facilitate ovulation in this species. Ovulation in *P. umbrina* seems to depend on a high rate of energy acquisition over a short period during early spring. In the wild, high food availability for *P. umbrina* is a reflection of the availability of water in shallow, ephemeral swamps in which, during spring, invertebrates and tadpoles – its main food – reach high densities. The water levels in these swamps depend on rainfall. *Pseudemydura umbrina* only lays eggs at the end of a good or average wet season. The trigger for ovulation is neither rain nor water depths nor moisture nor the general body condition of the female, but a feeding peak based on a high availability of food. It is unclear if this trigger for ovulation also operates through a body condition threshold (which would be higher than that for the initiation of vitellogenesis) or another physiological mechanism, since captive females have generally a higher body condition (more fat reserves) than wild ones, but still do not ovulate without a feeding peak. In regard to the initiation of vitellogenesis, *P. umbrina* can be compared to the giant tortoise which initiates an ovarian cycle in most years (income breeder). Regarding egg production, however, this species functions as a capital breeder: it only breeds if it can amass energy capital at a high rate just prior to the nesting season. In dry years, when this is not possible, *P. umbrina* aborts ovulation and skips reproduction instead of reducing egg numbers, as an income breeder like the Aldabra giant tortoise (Swingland and Coe 1978; Swingland and Lessells 1979) or the desert tortoise *Gopherus agassizii* (Henen 1997; Karl 1997) does. The desert box turtle *Terrapene ornata luteola* also appears to follow the capital breeder pattern of *P. umbrina* and abort ovulation during dry years (Nieuwolt-Dacanay 1997).

6.1.4 Social Factors and Population Density

In contrast to many lizards and snakes (Duvall et al. 1982; Licht 1984), there is little evidence that social interactions directly influence or trigger reproductive processes in chelonians. Dominance hierarchies in groups of captive chelonians (Harless 1979) sometimes influence access to food, which in turn influences body condition which may impact on reproduction, but it is doubtful that this plays a role in natural populations. If *Pseudemydura umbrina* males and females are kept together year-round in enclosures (as has been the case with the captive colony pre-1988), the larger males tend to chase females out of the water during the mating season (winter and early spring) and, thus, deny them or limit their access to food. Under these conditions, females do not ovulate and lay eggs (see Sect. 6.1.3). In the wild, *P. umbrina* disperses over wide, flooded areas during the mating season, and social interactions do not limit the food acquisition of females which, during that time, improve their body condition at a faster rate than males (King et al. 1998).

The effects of social factors on reproduction discussed above are artefacts of captivity: they are the result of keeping some or many individuals in a – compared to natural populations – very restricted area or under unnatural densities. Population densities in wild populations rarely reach levels where they limit the ability of individuals to access food. This rare situation occurs in some high-density populations of giant tortoises *Geochelone gigantea* on Aldabra atoll where food is limited and reproduction is tightly restricted by food supply (Swingland and Coe 1978). The tortoises also compete actively for shade, with bigger ones being more successful. Small females show a disproportionate mortality over large adult females, and females generally over males which are the larger sex (Swingland and Lessells 1979). In these populations, social interactions impact indirectly on reproduction, via differences in access to food and, therefore, body condition.

6.2 Endogenous Timing Mechanisms

The generation of a complete annual gonadal cycle through endogenous timing mechanisms, a complete circannual rhythm, requires that internal factors alone be capable of both inducing and terminating a gonadal cycle at approximately yearly intervals. Such an endogenous periodicity should be expressed without an external *zeitgeber*, under constant environmental conditions that are permissive to reproductive activity. A complete circannual rhythm has not been demonstrated yet for any chelonian species.

There are some inconsistencies and confusions with the usage of the terms "circadian" and "circannual" rhythms or fluctuations in the chelonian literature. Correctly, these terms describe endogenous, intrinsic rhythms, the phase lengths of which approximate but, generally, do not equal in the first instance 24 h and in the second instance 1 year ("*circa*" is a Latin adverb and means "approximately"). Circadian and circannual rhythms can only be studied under constant environmental conditions where they start to run free and where, without the respective *zeitgeber*, their phase lengths deviate from the astronomical phases (Gwinner 1986; Underwood 1992). Several chelonian papers use the terms "circadian" and "circannual" incorrectly as synonyms for "daily" and "annual" (e.g. Vivien-Roels et al. 1979; Choudhury et al. 1982; Mahapatra et al. 1986; Kim et al. 1987; Mahata-Mahapatra and Mahata 1991). The occurrence of daily or annual fluctuations of measured parameters, however, cannot and does not imply in itself the existence of endogenous circadian or circannual rhythms. Indications for endogenous, free-running circadian rhythms under constant laboratory conditions have, for example, been reported for the locomotor activity of *Testudo hermanni* (Thinès 1968), *Geochelone sulcata* (Cloudsley-Thompson 1970), *Gopherus polyphemus* (Gourley 1972, 1979) and *Trachemys scripta elegans* (Cloudsley-Thompson 1982) and for temperature preference in *T. s. elegans* (Jarling et al. 1989). Experiments by Hutton (1960) provided the only indication for a possible circannual rhythm in a chelonian under constant environmental conditions, concerning cyclic changes of food intake, blood lipoid phosphorus and blood amino acid nitrogen concentration in *T. s. elegans*.

Despite the fact that a complete circannual rhythm has not yet been convincingly demonstrated for any chelonian species, several experimental investigations on the effects of temperature and photoperiod on gonadal activity (see Sect. 6.1.1) revealed the involvement of endogenous timing mechanisms in the control of certain phases of chelonian gonadal cycles. In *Testudo hermanni*, the onset of the spermatogenetic cycle is partly under endogenous control: the reaction-readiness of the hypothalamus–pituitary–gonad axis increased endogenously during March so that gonial proliferation started spontaneously under hibernation conditions (5 °C and constant darkness). Although spermatogenesis is primarily under temperature control, spermatocytogenesis also commenced spontaneously (although at a very slow rate) if hibernation conditions were extended until June. The termination of spermatogenesis also required an endogenous readiness of the hypothalamus–pituitary–gonad system to react upon low temperature: from June to August a constant temperature of 5 °C and constant darkness did not induce regression, and spermatocytogenesis continued strongly, whereas the same treatment during September and November caused immediate regression of the testes. However, spermatogenesis continued under warm conditions until November and January; total regression of the testes was always correlated with low environmental temperatures (Kuchling 1982a). In *Kinosternon odoratum*, testicular regression also depended partly on an endogenous readiness and an endogenous mechanism may have affected peak androgen secretion (Mendonça and Licht 1986b).

In many lizards and snakes, gonadal quiescence in the post-breeding season results from a refractory condition whereby the animal is rendered insensitive and unresponsive to stimulation by exogenous conditions that would otherwise represent optimal conditions for gonadal activity. Evidence is accumulating that gonadal refractoriness in lizards and snakes is to a large extent controlled by endogenous physiological rhythms (Fischer 1974; Licht 1984; Underwood 1992). Ganzhorn and Licht (1983) speculated that testis regression in *Chrysemys picta* also may, at least partly, be controlled by the development of an endogenous refractoriness to stimulatory thermal conditions. These authors applied the term "refractoriness" in a very wide sense, even to a slight reduction in the intensity of spermatogenesis: although testes weights and plasma testosterone concentrations declined in autumn under warm environmental conditions, the histological data still suggested active spermatogenesis. Since spermatogenesis was not blocked in *C. picta*, I would call this a reduction of the endogenous readiness of the testes to stimulating environmental conditions rather than refractoriness. Refractoriness, however, is not a clearly defined concept and its use is arbitrary, particularly in the literature on reptiles. Botte and Angelini (1980) defined the refractory period of reptiles as an interruption of gonadal function, plus an insensitivity of the gonads to stimulatory environmental conditions, as opposed to a simple slowing down of gonadal activity. In the same paper, however, these authors diluted again their own concept of refractoriness by applying the term refractory period also to the time when the "slowing down of gonadal activity becomes more intense and synchronised in all adult populations" (Botte and Angelini 1980: 204). Underwood (1992: 275) considers post-breeding refractoriness to occur when "the reproductive system collapses under conditions of light or temperature that are normally stimulatory to it". This collapse was certainly not evident with the testes in the experiments on *C. picta* described by Ganzhorn and Licht (1983).

In any case, an endogenous component seems to be involved in testicular regression of chelonians: in *Testudo hermanni*, testicular regression is primarily under the control of environmental temperature, but an underlying readiness to react to the declining temperature is also necessary (Kuchling 1982a). However, since stimulating environmental conditions easily override the time programme for regression, I do not call this phenomenon refractoriness. In *Chrysemys picta* females, Ganzhorn and Licht (1983) suggested that the seasonal differences in responsiveness of the ovaries to environmental stimuli are also partly related to some endogenous cycle in their sensitivity to exogenous factors. All these experimental results with temperate-zone chelonians suggest the possibility, but do not prove, that an endogenous, circannual rhythm or a circannual clock may underlie chelonian gonadal cycles.

Due to their phylogenetic placement (see Sect. 1.1), chelonians may provide excellent models to gain insights into the evolution of refractoriness and circannual rhythmicity through a comparative approach. Chelonians also offer a variety of natural experiments or "experiments of nature", since several taxonomic groups with highly seasonal reproduction occur in equatorial areas. Other groups show distributions from equatorial to subtropical or temperate areas. An endogenous circannual reproductive rhythm would seem particularly adaptive in equatorial rain forest species which show strongly seasonal breeding cycles under more or less constant conditions of light, temperature and moisture, for example some river turtles. Although falling or low water levels may provide important timing cues for nesting (see Sect. 6.1.2), their gonadal cycles have to start long before these conditions prevail in order for the reproductive system to be ready at the "proper" time of the year.

It is, therefore, not surprising that the best supportive evidence for an annual refractory period in any chelonian comes from a study of the regulation of ovarian activity in a tropical (although not equatorial) species, the soft-shelled turtle *Lissemys punctata*: during the quiescent phase, neither ovarian growth nor steroidogenesis can be stimulated by various photothermal treatments (Sarkar et al. 1996a). However, Sarkar et al. (1996a) consider a lack of food intake during that time responsible for the complete shutdown of ovarian responsiveness to environmental stimuli rather than an involvement of post-breeding refractoriness and/or an endogenous component, which, of course, may be the case, since proving an endogenous mechanism requires rigorously controlled experiments. However, the experimental results with *L. punctata* support the hypothesis that tropical chelonians with seasonal reproductive cycles may be good candidates for the study of endogenous circannual rhythms.

There is also circumstantial evidence that at least some temperate-zone chelonians should be able to measure the passage of time endogenously, but these species have not been experimentally tested. These are species which show long periods of aestivation and/or hibernation in situations where environmental cues are very limited. In *Pseudemydura umbrina* for example, the ovarian cycle – the vitellogenic growth phase of the follicles – commences during aestivation in January or February (Kuchling and Bradshaw 1993). From November or early December until April or May these turtles do not feed and remain deep in underground tunnels or buried under leaf litter and soil (Burbidge 1981), refuges that provide few light or temperature cues. *Chelodina steindachneri* may be buried underground for unpredictable periods at any time of the year (see Sect. 5.4), yet their testicular and ovarian cycles commence at the appropriate time (G. Kuchling and S.D. Bradshaw, in prep.). It seems likely that, in chelonians which show prolonged periods of dormancy, selection may also favour the development of endogenous circannual rhythmicity.

Age and body size are other intrinsic factors which influence onset and duration of gonadal activity. Older and larger males may begin spermatogenesis earlier and continue it longer in the year than younger and smaller ones. With respect to females, older and larger females may also nest earlier and, in species with multiple clutches, continue nesting later than younger and smaller females of the population (Moll 1979).

6.3 Effects of Stress on Reproduction

Stress, for example the stress of captivity (see Sect. 2.5), can clearly impact on chelonian reproduction. Observations on stressed captive *Chelydra serpentina* suggest that the response of reproductive processes to stress varies in intensity according to the state of the testicular or ovarian cycle and the condition of the gonads. It is easier to block gonadal growth at the start of a new gonadal cycle than to inhibit or interrupt spermatogenesis or vitellogenesis once it has begun. *Chelydra serpentina* females abort ovulation if caught in spring prior to ovulation and kept in the holding tanks of a commercial trader, although females collected in autumn with preovulatory follicles and maintained in university holding facilities ovulate in spring (Mahmoud and Licht 1997). These results suggest that some of the environmental effects on reproduction, in particular in extreme and/or unpredictable environments (see Sect. 5.4), could simply be interpreted as effects of environmental stress.

The question of stress is difficult to deal with, and there is no general consensus on how to define stress. It is, therefore, not easy to establish if and when an animal in its natural environment is stressed or to measure the effects of stress on animals. Although various concepts of stress have been proposed, from an eco-physiological point stress is best defined as "the physiological resultants of de-mands which exceed an organism's regulatory capacities" (Bradshaw 1997: 154). A stressed individual would thus be one showing evidence of an activated, but inadequate, regulatory response to change. This concept allows us to dissociate potentially harmful changes of state from beneficial regulatory responses: an environmental change is not considered a stressor as long as the regulatory response of the animal is adequate to maintain its normal functions. The defini-tion of stress is not only an academic question but also a practical and important one for conservation management, since environmental stress can be involved in the decline of populations and is often part of the threats to the survival of species. However, due to the complexity of the reproductive processes, it is often difficult to judge what constitutes an inadequate regulatory response in regard to reproduction.

In late 1990, the last surviving wild population of the critically endangered western swamp tortoise *Pseudemydura umbrina* was protected by the con-struction of a predator-proof fence around its habitat of 29 ha (Kuchling et al. 1992). Part of the home range of an old *P. umbrina* female (>50 years old) was obviously located outside the nature reserve and cut off by the new fence. This female, which had reproduced in 1989 and 1990, spent much time trying to get through the fence and did not lay eggs in 1991 and 1992 despite showing normal vitellogenetic cycles, and despite having preovulatory follicles at the start of the breeding season. Since the other monitored females in the nature reserve did lay eggs in 1991 and 1992, the new fence was, obviously, the reason why this female failed to breed. In late 1992 she was transferred to the captive colony and, since then, has been producing eggs on an annual basis. For this particular female (which, then, represented about 10% of the wild world population of adult *P. umbrina* females), the predator-proof fence was a stressor which caused her failure to reproduce, which would have been part of maintaining her normal functions.

However, if *P. umbrina* females or *Chelodina steindachneri* females do not ovulate and produce eggs due to dry conditions which are part of natural environ-mental fluctuations (see Sect. 5.4), the strategy of not breeding in those situations is adaptive, part of their regulatory capacity to maintain their normal functions in their habitat and, therefore, not a sign of stress according to Bradshaw's (1997) operational stress concept. The same environmental conditions would be stres-sors if females were maladapted, bred and then lost body condition and/or risk dying through desiccation during aestivation. In the ancestors of *P. umbrina* and *C. steindachneri*, exactly this environmental stress must have played a critical role in the evolution of the novel adaptation – to abort or delay ovulation and ovipo-sition under conditions of drought. Stress, thus, is perceived as a temporary and maladapted state existing between homeostasis (or normal functioning) and pa-thology. Some challenge through environmental "stress" seems necessary for a species to react in an evolutionary way to environmental changes (Bradshaw 1997).

Species-specific differences in the susceptibility of turtles to stress are certainly widespread. The multi-clutch sea turtles, for example, typically ovulate the next clutch of eggs within a day or two after nesting. *Caretta caretta* ovulated when, after nesting, it was held in a crowded shallow tank for 3 days (Wibbels et al. 1992). *Chelonia mydas* ovulated even when, after nesting, it was flipped over and left on its back on the beach (Licht et al. 1980). However, post-nesting female *Lepidochelys olivacea* did not ovulate when restrained in a shaded corral on the beach, although unrestrained post-nesters captured in the water the day after nesting had ovulated (Owens 1997). In these situations, *Lepidochelys olivacea* was clearly stressed in the sense of Bradshaw's restricted concept of stress, and seemed to be more prone to stress from confinement and captivity than the other two species.

This rigorous use of the term stress contrasts markedly with the common practice in the chelonian literature to assume that any manipulation which causes a measurable change in an animal causes stress. For example, several studies found that plasma concentrations of gonadal steroids drop within a few hours of the capture of various wild chelonians (*Chrysemys picta*: Licht et al. 1985a; *Kinosternon odoratum*: Mendonça and Licht 1986a; *Chelydra serpentina*: Mahmoud et al. 1989; Mahmoud and Licht 1997), and that plasma levels of corticosteroids increase (*Lissemys punctata*: Mahapatra et al. 1991; *Caretta caretta*: Gregory et al. 1996) – phenomena which these authors attribute to stress. Although it clearly is a nuisance for the design of endocrinological studies and for the evaluation of earlier investigations that the transfer of wild animals into captivity for a few hours or days alters the hormonal levels from the natural conditions, these changes may simply be adequate regulatory responses. Plasma sex steroid levels in *Chelodina oblonga* males (Fig. 6.1) and females (Figs. 6.2, 6.3) decrease even if the turtles are maintained in wire mesh enclosures directly in the lake where they were caught. However, these hormonal changes due to handling or short-term confinement may be part of the animal's normal response to disturbance and, if the disturbance does not persist, may not lead to a significant alteration of its normal functioning (e.g. gamete production and/or breeding). Due to Bradshaw's (1997) stress concept, a short-term captive animal with reduced gonadal steroid levels would only be considered to be stressed if also experiencing a significant level of disturbance of its reproductive functions. If the same disturbance, e.g. captivity, becomes chronic, it may well cause stress: the transfer of wild chelonians into laboratories or outdoor enclosures can seriously disturb or alter reproductive processes (see Sect. 2.5).

6.4 Conclusions

Seasonal reproduction in chelonians is controlled by a combination of endogenous and exogenous factors, the relative importance of which varies among cycle phases and species. The limited experimental data which are available on the control of reproduction in chelonians indicate that the initiation of gonadal cycles, which typically takes place half to three quarters of a year before the actual breeding season, involves important endogenous or pre-programmed control mechanisms in all species that have been studied. Intensity and termination of gonadal activities depend heavily on environmental factors, with temperature

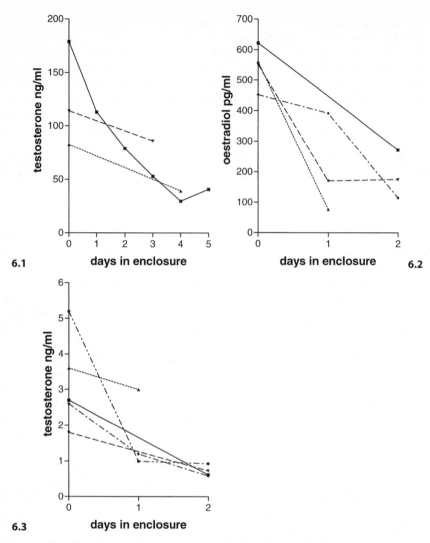

Fig. 6.1. *Chelodina oblonga.* Plasma testosterone levels of three males during spermiation, kept for up to 5 days in a wire mesh enclosure of 3 m diameter in the lake where they were caught
Fig. 6.2. *Chelodina oblonga.* Plasma oestradiol levels of four females during late vitellogenesis, kept for up to 2 days in a wire mesh enclosure of 3 m diameter in the lake where they were caught
Fig. 6.3. *Chelodina oblonga.* Plasma testosterone levels of five females during late vitellogenesis, kept for up to 2 days in a wire mesh enclosure of 3 m diameter in the lake where they were caught

being the one of foremost importance. Ovulation and oviposition – energetically the most critical phases of reproduction – seem to rely more heavily on pre-programmed mechanisms (closed control) in species occupying predictable environments than in species living in unpredictable habitats in which exogenous cues become more important (labile control *sensu* Whittier and Crews 1987).

These labile control mechanisms either retard reproductive processes (gravidity and/or oviposition) until conditions are favourable or, in bad years, may totally abort them. Once a new gonadal cycle has begun, stress effects (e.g. captivity) on reproduction seem to be less pronounced in species with primarily pre-programmed control mechanisms than in species with labile control mechanisms.

Chapter 7

Hormones and Reproduction

Exogenous and endogenous regulators of reproductive cycles (Chap. 6) operate through two integrative systems, the nervous system and the endocrine system, which also coordinate the different components of reproduction with each other. The nervous system conducts electrical impulses at great speed and, generally, controls rapid processes. The endocrine system regulates long-term processes within the animal by building up and sustaining levels of hormones in the blood which bind to target organs and trigger physiological responses. Hormones play an essential role in the control and integration of the complex processes of reproduction. The nervous system co-operates in this task. The neuroendocrine mechanisms are correlated with the environment and are central to, and finely tuned by, complex behavioural interactions.

Two types of hormones, the neurohormones and the glandular hormones, are directly involved in the regulation of reproduction in vertebrates. A third type of hormones, the tissue hormones, have no direct role in the regulation of reproduction. Neurohormones are formed in modified nerve cells called neurosecretory neurones which receive signals from "normal" nerve cells. The glandular hormones are produced in specialised endocrine glands. Hormones are chemically heterogeneous, with those of vertebrates falling into three main groups: the polypeptide or proteohormones, the steroid hormones and amino acid derivatives. The first group consists of polypeptides or relatively small proteins with a maximal molecular weight of about 40 kDa. The steroid hormones, the second group, all have a characteristic chemical structure which derives from the saturated tetracyclic form of cyclopentanoperhydrophenanthrene (or gonan). All male and female sex hormones are steroids. The third hormone group of amino acid derivatives has a low molecular weight, for example adrenalin and the thyroid hormones (Blüm 1986).

Studies of the reproductive endocrinology of chelonians commenced in the mid-1930s when Risley (1937) studied the effects of testosterone on sex development in turtle embryos and the effects of gonadotropic and sex hormones on juvenile diamond-back terrapins *Malaclemys terrapin* (Risley 1939). In the 1950s, the effect of hypophysectomy on testicular structures was investigated in the freshwater turtle *Mauremys leprosa* (Combescot 1955b) and *Testudo graeca* (Combescot 1955c), but these studies were very basic and provided only very limited insights. It was not until the 1970s that endocrinological studies of chelonians were intensified when fresh pituitary glands of hundreds of green turtles *Chelonia mydas*, which were commercially farmed at the Grand Cayman Island, were used to extract and purify their peptide hormones. Their structural and biological properties were then compared with those of many other chelonian and vertebrate species (for review see Licht 1983). Also in the 1970s, complete annual cycles of the plasma concentrations of gonadal steroids were for the first time measured quantitatively in chelonians by radioimmunoassay (Callard et al. 1976, 1978; Kuchling 1979; Lewis et al. 1979; Kuchling et al. 1981). Since then, a number of laboratory investigations and hormone studies of wild populations have further improved our knowledge of the endocrinology of chelonians.

7.1 Gonadal Steroids

The main role of the gonads is the production of gametes (Chap. 3), but, parallel to and integrated with this process, gonads also have an important endocrine function. The testes produce primarily the male sex hormones (androgens) and the ovaries the female sex hormones (oestrogens and progesterone, or gestagen, but also androgens). These hormones influence a range of reproductive phenomena, for example the differentiation of the embryonal gonads (see Sect. 8.3), the development of primary and secondary sexual characteristics (see Sect. 1.4), the functional preparation of the Wolffian (epididymis and vas deferens) and Müllerian ducts (oviducts) to accommodate and process the gametes (see Sects. 3.2.5, 3.3.6), the production of scents and attractants, and reproductive behaviour (Chap. 4). According to the various functions, the secretion of sex steroid hormones shows annual cycles which are related to, but not necessarily parallel to, the cycles of gamete production.

7.1.1 Males

Studies of annual cycles of steroidogenic activity in the chelonian testis commenced in the 1970s. The first approaches relied heavily on histological and histochemical techniques of presumed steroidogenic tissues, mainly the interstitial Leydig cells and, within the tubuli seminiferi, the Sertoli cells. Callard et al. (1976) combined these techniques with the direct measurement of plasma testosterone concentrations by radioimmunoassay in *Chrysemys picta* and found that testosterone levels are highest in winter and spring, when Leydig cells are large and the germinal epithelium is inactive, and lowest, at a basal level, during active spermatogenesis in summer and autumn. Based on histological criteria of cell volume, nuclear diameter and lipid content, Lofts and Tsui (1977) described a similar pattern for the soft-shelled turtle *Pelodiscus sinensis*, with endocrine cell activity in the testis at a peak during gametogenic quiescence, decreasing as spermatogenesis is renewed and minimal at the peak of spermiogenesis. In the late 1970s, this pattern was regarded as typical for chelonians (Moll 1979). Callard and Ho (1980) and Lance and Callard (1980) proposed a generalised androgen cycle for the male turtle in which plasma testosterone peaks during the time of gametogenic quiescence in spring (March/April), decreases dramatically during

May, is at a low baseline level during summer when spermiogenesis peaks and increases again slowly during autumn (Fig. 7.1). This concept assumes that mating activity occurs when Leydig cells are most active and peripheral androgen is at a peak in early spring, whereas the period of active spermatogenesis is associated with atrophic Leydig cells and low plasma androgen.

The first challenge to the accepted wisdom of that time came from my study of the testicular and androgen cycle of the European tortoise *Testudo hermanni*, the results of which are not consistent with the generalised androgen cycle proposed by Callard and Ho (1980) and Lance and Callard (1980). In *T. hermanni*, plasma androgen levels closely follow testicular growth and spermatogenesis and peak during spermiogenesis (Fig. 7.2; Kuchling 1979; Kuchling et al. 1981). The discovery of this pattern – basically the opposite of Callard and Ho's (1980) concept of a separation of steroid secretion and spermatogenesis – generated a rush of discussions of similar, unpublished data for several other turtle species. Obviously, data which conflicted with the then accepted concept had been available in

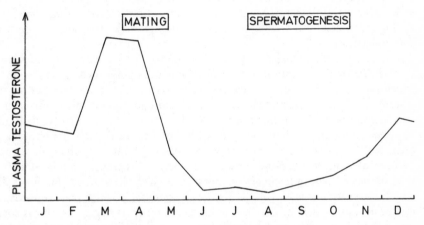

Fig. 7.1. Plasma testosterone concentration, spermatogenesis and mating of seasonally breeding male turtles (according to Callard and Ho 1980; Lance and Callard 1980)

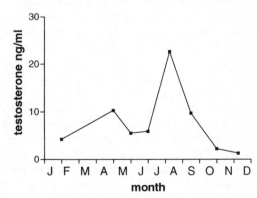

Fig. 7.2. *Testudo hermanni.* Changes in plasma testosterone concentration of males in a population in Montenegro, former Yugoslavia. (Redrawn from Kuchling et al. 1981)

125

some laboratories, but, up to that point, had not been published. Based on unpublished observations and data, Licht (1982), in a review of endocrine patterns of reproductive cycles in turtles, figured graphs of annual plasma testosterone levels of *Pelodiscus sinensis* and *Kinosternon odoratum* which grossly corresponded to the pattern of *T. hermanni* (Fig. 7.2). In the same year, the publication of plasma testosterone data of *K. odoratum* confirmed this pattern for this species (McPherson et al. 1982). *Pelodiscus sinensis* is one of the species on which the original concept of Callard and Ho (1980) had been developed, which illustrates the errors that can result from reliance upon histology as a criterion of steroidogenesis. It is now well established that the concept first presented by Kuchling (1979) and Kuchling et al. (1981) is the most common pattern of androgen secretion in male chelonians. Loggerhead sea turtles *Caretta caretta* also show high testosterone levels during spermiogenesis, which decline prior to the mating season. A distinct dichotomy in testosterone levels in males of that species reflects their multi-year testicular cycles (see Sect. 5.3; Wibbels et al. in Owens 1997). *Chrysemys picta*, however, seems to be an exception. For this species, Licht et al. (1985a) corroborated to some degree the earlier results of Callard et al. (1976), with a plasma testosterone peak during the mating time in early spring, a low level during summer and a second peak in autumn, after the late summer peak in spermiogenesis.

Although the phenological description of annual androgen cycles is interesting in itself, the major question is their physiological significance, the question of which reproductive processes they actually regulate. A standard concept for the postnuptial type of spermatogenesis is that androgens peak during the mating season and that high plasma androgen levels are related to mating (Callard and Ho 1980; Lance and Callard 1980; Licht 1984; Mesner et al. 1993; Mahmoud and Licht 1997). This is, in fact, the case in the spring-mating *C. picta* (Callard et al. 1976; Licht et al. 1985a). However, in other species this relationship is either less clear or may even be opposite. For many species, the exact timing of mating is still largely unresolved (see Sect. 4.2.1). *Testudo hermanni*, in which plasma testosterone peaks during maximal testis development and spermiogenesis, shows some mating activity at any time during their activity period, although this species does show peak mating activity in spring (Kuchling 1982a) at a time when plasma testosterone is slightly elevated, but not at a maximal level (Kuchling et al. 1981). *Trachemys dorbigni* in southern Brazil shows a single plasma testosterone peak at the time of spermiogenesis and a low, basal level during the mating time in spring (Silva et al. 1984). In *K. odoratum*, plasma testosterone levels in spring are also at a low, basal level and show a single peak during spermiation in autumn (Mendonça and Licht 1986a). However, *K. odoratum* seems to have its main mating period in autumn, but may also mate in spring (Mendonça 1987b). *Gopherus agassizii* also shows plasma testosterone at a high level during the autumn mating period, but low, basal levels during the spring mating period (Rostal et al. 1994). In the sea turtles *Chelonia mydas* (Licht et al. 1979) and *Caretta caretta* (Wibbels et al. 1990), plasma testosterone drops between the prebreeding season and the height of the copulatory period. Although the administration of testosterone did lead to courtship and mating attempts in immature male *C. mydas*, a correlation exists in male and female sea turtles between high

126

testosterone levels and migration behaviour rather than with mating (Owens 1997). In most chelonians, no clear or coherent correlation exists between high plasma testosterone and mating activity, and the attempts of many authors to construe one seem rather awkward.

Kuchling et al. (1981) interpreted the high plasma testosterone level in *T. hermanni* in summer in relation to high enzymatic activity in the epididymis and related maturation processes of spermatozoa. In *C. picta*, the cytoplasmatic droplets of all spermatozoa which are stored in the epididymis become detached from the sperm midpiece in a synchronised and coordinated manner during autumn (Gist et al. 1992), an event which may well be triggered by the elevated androgen levels found in this species during autumn (Licht et al. 1985a). Sperm maturation in the chelonian epididymis – a prerequisite for sperm to become ready for fertilisation – involves enzymatic processes which seem to depend on the secretory activity of the epididymis (see Sect. 3.3.6) and which, in turn, seem to be androgen dependent. However, the phenomenon of sperm maturation in the epididymis has so far largely escaped the attention of most chelonian investigators. Its timing may potentially be correlated to some of the observed inconsistencies of seasonal androgen levels between chelonian species.

An enzyme involved in the synthesis of androgens is the 3β-hydroxysteroid dehydrogenase. Histochemical data for various chelonians show a temporal separation of 3β-hydroxysteroid dehydrogenase activity between the extratubular Leydig cells and the Sertoli cells of the germinal epithelium during the testis cycle (*P. sinensis*: Lofts and Tsui 1977; *T. hermanni*: Kuchling et al. 1981; *Chelydra serpentina*: Mahmoud et al. 1985; Mahmoud and Licht 1997; *C. picta*: Dubois et al. 1988). Cycles of Leydig cells and epididymis mass are essentially inverse to testis mass and spermatogenetic cycles. The direct, separate measurement of androgen synthesis in the Leydig cells and Sertoli cells of *C. serpentina* confirmed that testosterone concentration in the Leydig cells peaks during the mating period in May and early June, whereas intratubular testosterone peaks during maximal testicular development with spermiogenesis and spermiation; at that time, the Leydig cells show a second, lower peak; when testes regress in autumn, testosterone from both sources declines significantly (Mesner et al. 1993). Circulating androgens either originate from the interstitial Leydig cells or from the intratubular Sertoli cells. Mesner et al. (1993) demonstrated that, throughout the cycle, the Leydig cells produce more testosterone per milligram of tissue protein than the Sertoli cells, and concluded that the Leydig cells are the major site of testosterone synthesis. However, regarding the overall contribution of those compartments to plasma testosterone, the hugely expanded tubules during spermiogenesis may still release quantitatively significant amounts of androgens into circulation. In *T. hermanni*, the smaller peak of plasma testosterone during the spring mating season and the main peak during spermiogenesis in late summer (Fig. 7.2) may reflect a biphasic pattern of androgen secretion in the testis comparable to that of *C. serpentina*.

Circulating steroid levels may not necessarily reflect only steroid secretion rates of the testes, they can also be influenced by turnover and clearance rates which could be altered by changes in body temperature. A rise in plasma steroid levels during cold periods (as, for example, described in *C. picta* by Callard et al. 1976) could reflect low turnover or clearance rather than increased gonadal activity. In *T. hermanni*, circulating testosterone levels are modulated by long-term changes of the environmental temperature, with the height of the testosterone levels depending at least in part on the height of the temperature (Vivien-Roels and Arendt 1981; Kuchling 1982b). Behavioural thermoregulation also influences testosterone levels. In an experiment during autumn, males of *T. hermanni* which, through behavioural thermoregulation, reached higher maximal body temperatures during the day also had significantly higher testosterone levels than those tortoises which remained at a lower maximal temperature (Fig. 7.3; Kuchling 1986). *Chelydra serpentina* males, kept for 5 days in tanks at various temperatures at the start of spermiogenesis in July, showed significantly lower declines in testosterone levels when kept inside their thermoactive range (20–26 °C) than when kept at temperatures either above (29–32 °C) or below (14–17 °C) this range (Mahmoud and Licht 1997). These observations suggest that variations in steroid secretion and release rather than clearance rates are the principal determinants of peripheral testosterone levels in *T. hermanni* and *C. serpentina*.

Only few data are available on the release pattern of androgens into the bloodstream. Individuals tend to show daily fluctuations of plasma androgens, but, due to high variability of concentrations among individuals, clear daily cycles are

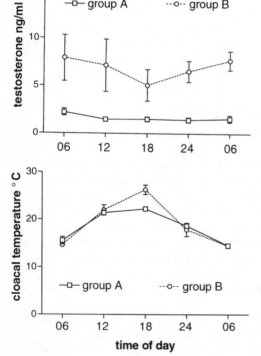

Fig. 7.3. *Testudo hermanni* males. Changes of cloacal temperature and plasma testosterone concentration over 24 h; time of spermiation, towards end of activity period in October. The group which reaches a higher maximal cloacal temperature during the day shows higher plasma testosterone levels. (Data from Kuchling 1986)

often not detectable when populations are analysed. Twenty-four hour plasma testosterone levels of individual males of *Chelonia mydas* and *T. hermanni* fluctuate in a rather inconsistent way. In *C. mydas*, there was a hint for a nocturnal peak during September but not in other seasons (Licht et al. 1985b), whereas *T. hermanni* tends to show a peak in the late afternoon during May (mating season) and a peak in the early morning during August (spermiogenesis) and October (Kuchling 1986). Just before the start of their mating season in July, individual males of *Chelodina oblonga* show a diurnal peak of plasma testosterone at midday and low levels during the night, with the fluctuations roughly paralleling those of the cloacal temperatures (Fig. 7.4). In the investigations of all three species, individual plasma testosterone concentrations were consistently lower at the end than at the start of the 24-h sampling periods, which can be attributed to the effects of handling and repeated sampling (see Sect. 6.3).

The early studies of plasma steroid cycles, including those of *T. hermanni* (Kuchling et al. 1981), *P. sinensis* (Licht 1982), *K. odoratum* (McPherson et al.

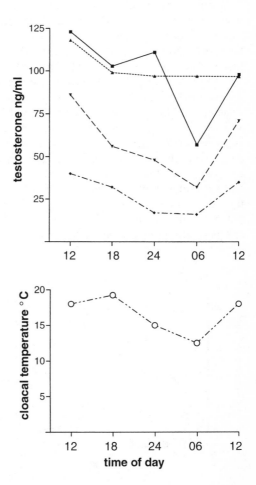

Fig. 7.4. *Chelodina oblonga* males. Changes of cloacal temperature and individual plasma testosterone concentrations over 24h; time of spermiation, early July

1982; Mendonça and Licht 1986a) and *T. dorbigni* (Silva et al. 1984), show relatively sharp peaks of plasma testosterone concentration at the time of maximal spermiogenesis. In those studies, however, blood samples were not obtained immediately from freshly captured wild specimens. All these studies were conducted on animals held in captivity from a few hours to several days, which may have altered their "normal" hormonal levels (see Sects. 2.5 and 6.3). Mahmoud and Licht (1997) demonstrated that, if *C. serpentina* are maintained inside their thermoactive range, testosterone levels show a lower decline in response to short-term captivity and periodic bleeding at the time of spermiogenesis in summer than they do during spermatocytogenesis in spring. If a similar situation occurs in other turtle species, this phenomenon may have exaggerated the sharp peaks of plasma testosterone during spermiogenesis – relative to androgen levels which at other times decline more quickly – in the studies mentioned above. The sharp testosterone peaks may be artefacts owing to the fact that, under short-term captivity, testosterone is more resilient during spermiogenesis when the peaks were observed, but declines quickly at other times.

Comprehensive data from freshly captured wild turtles became available only recently and show elevated plateaus of plasma testosterone over extended periods (several months) rather than peaks, e.g. in male *C. serpentina* (Mahmoud and Licht 1997), in male sea turtles (Owens 1997) and in males of the Australian chelid turtles *Chelodina oblonga* (Fig. 7.5) and *C. steindachneri* (G. Kuchling and S.D. Bradshaw, in prep.). Mahmoud and Licht (1997) interpreted the extended testosterone plateau of *C. serpentina* in relation to the short "compressed" testis cycle of their study population from the northern temperate region (Wisconsin), and in relation to the observed non-seasonality of mating in that population. However, *C. oblonga* and *C. steindachneri*, despite showing long testicular cycles and despite showing their main mating activity during winter when testosterone levels are low, also exhibit a long plateau of plasma testosterone during summer and autumn, relatively similar to that of *C. serpentina*. Rather than being specific for cool temperate regions, this pattern may be typical for most species, but may only become evident if freshly captured wild turtles are studied. An alternative hypothesis would be that the family Chelidae (*C. oblonga* and *C.*

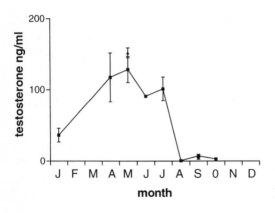

Fig. 7.5. *Chelodina oblonga* males. Changes in plasma testosterone concentration of males in a population near Perth, Western Australia

steindachneri, suborder Pleurodira) has an androgen pattern similar to the Chelydridae and Cheloniidae (suborder Cryptodira), but the cryptodire families Emydidae, Kinosternidae, Testudinidae and Trionychidae have different patterns. In any case, it would be interesting to re-study some of the species for which sharp peaks of circulating testosterone were reported, in order to find out if they show plateau levels of circulating testosterone when freshly captured wild turtles are sampled. This hypothesis, however, still cannot account for the marked differences in observed testosterone patterns between male *C. picta* and other turtle species.

7.1.2 Females

The most extensively studied reproductive events of chelonians are certainly the nesting and egg laying of sea turtles. Female sea turtles, when crawling, digging and ovipositing on beaches, are very obvious and easy to access, a situation which rarely applies to other chelonians. Due to their size and the relative ease of blood sampling (see Sect. 2.4), the reproductive endocrinology of adult female sea turtles of the family Cheloniidae is relatively well researched. According to a general model proposed by Owens (1997), plasma levels of all sex steroids are low in reproductively inactive periods of the multi-year cycles. Oestrogen peaks and testosterone increases slightly during vitellogenesis when turtles are in their foraging grounds. Just before the onset of their migration, oestrogen decreases and testosterone increases and, possibly, reaches its highest level in receptive females during the mating period which precedes ovulation of the first clutch. Progesterone increases slightly during the mating time and shows dramatic surges at the time of the ovulation of each of the multiple clutches, preceded by slight elevations of both oestrogen and testosterone which decline abruptly once progesterone surges. All sex steroids decrease during the return migration to the foraging grounds (Fig. 7.6). This seems to be a good general working model for sea turtles, although some species-specific differences exist (Wibbels et al. 1992). Over the course of the nesting season, the leatherback sea turtle *Dermochelys coriacea* (family Dermochelyidae) shows similar patterns of plasma steroids, but circulating levels of testosterone and oestradiol are overall higher than in other sea turtles (Rostal et al. 1996).

This model also corroborates and corresponds well to data on annual cycles of ovarian steroids in the freshwater turtles *Chrysemys picta* (Callard et al. 1978), *Chelydra serpentina* (Lewis et al. 1979; Mahmoud and Licht 1997), *Kinosternon odoratum* (McPherson et al. 1982) and *Lissemys punctata* (Sarkar et al. 1996b). In all these species, oestrogen levels are elevated during vitellogenesis when compared to the basal levels during ovarian inactivity. Oestrogen and progesterone concentrations increase during the periovulatory period, with the oestrogen peak preceding that of progesterone, and decline as shelling of the eggs progresses in the oviducts. Testosterone (reported for *C. picta*, *K. odoratum* and *C. serpentina*) increases slightly at the onset of vitellogenesis in autumn and, with the exception of *C. serpentina*, also peaks during the periovulatory period. In the multi-clutched

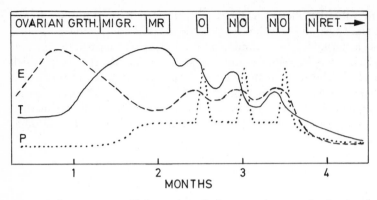

Fig. 7.6. Plasma sex steroid fluctuations during an active reproductive season of female sea turtles. *E* oestrogen; *MIGR* migration; *MR* mating receptivity; *N* nesting; *O* ovulation; *P* progesterone; *RET* return migration; *T* testosterone. (Adapted from Owens 1997)

K. odoratum, plasma testosterone is only elevated prior to ovulation of the first clutch, but not prior to subsequent ovulations (McPherson et al. 1982). Surges in plasma testosterone also occur in desert tortoise *Gopherus agassizii* females with preovulatory follicles (Rostal et al. 1994) and in females of *Chelodina steindachneri* with preovulatory follicles (Fig. 7.7). *Chelodina steindachneri*, however, does not show a periovulatory peak in plasma oestradiol 17-β which seems to circulate at lower levels than it does in other freshwater turtles (Fig. 7.7; G. Kuchling and S.D. Bradshaw, in prep.), which raises the question of whether another oestrogen may be present that is not detected by the specific radioimmunoassay. However, the green sea turtle *Chelonia mydas* also shows consistently lower oestrogen levels than other sea turtles (Wibbels et al. 1992) and this phenomenon, although difficult to explain, may be species-specific. Relatively low levels of oestrogen may be sufficient in these species to support vitellogenesis.

During nesting of *Caretta caretta*, plasma oestrogen is at an undetectably low level. Progesterone, testosterone and corticosterone profiles are associated with the progression of nesting episodes, with all three hormones declining between successive nestings and at a very low level by the last nesting episode. The magnitude of change between testosterone and corticosterone is highly correlated, suggesting an interaction of these two hormones which may be important in regulating ovarian and hepatic functions in regard to the mobilisation of reserves for vitellogenesis of subsequent clutches (Whittier et al. 1997).

From a functional perspective, oestrogen is the primary stimulus for vitellogenesis in reptiles (Ho 1987), including turtles (Heck et al. 1997). Progesterone may be involved in oocyte maturation including germinal vesicle breakdown prior to ovulation (Nagahama 1987), egg white protein production (O'Malley et al. 1969) and control of oviducal motility (Callard and Hirsch 1976; Mahmoud et al. 1988; Giannoukos and Callard 1996). In contrast to mammals and birds, oestradiol increases both the oestrogen and progesterone receptors in the oviduct of *Chrysemys picta* and progesterone downregulates both receptors (Giannoukos

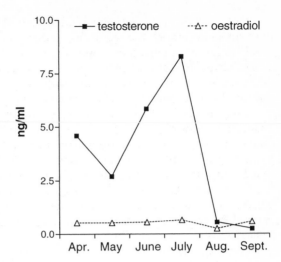

Fig. 7.7. *Chelodina steindachneri* females. Changes in plasma concentrations of oestradiol and testosterone during late vitellogenesis (April–July), ovulation (August) and oviducal period of eggs (September)

and Callard 1996). The specific action of testosterone during early vitellogenesis and during the preovulatory period in chelonians has never been experimentally addressed. Testosterone could sensitise preovulatory follicles to luteinising hormone stimulation (Wibbels et al. 1992), play a role in the nesting physiology and in the regulation of vitellogenesis (Whittier et al. 1997) and/or regulate migration behaviour and receptivity of females (Owens 1997).

7.2 Gonadotropins

The gonadotropin system of tetrapods, with the exception of some squamate reptiles, has two components which affect the gonads and which are synthetised and released by the adenohypophysis: follicle stimulating hormone (FSH) and luteinising hormone (LH). Another hormone of the adenohypophysis, prolactin, is also of direct relevance to reproduction, particularly reproductive behaviour, but strictly speaking it is not a gonadotropin. Thyrotropin, also a hormone of the adenohypophysis, regulates thyroid hormones which have some indirect involvement in the regulation of reproduction. These hormones will be discussed in Sections 7.4 and 7.5.

Gonadotropins are glycoprotein hormones which consist of an α and a β subunit, with the β subunit determining the biological activity of the hormone. In contrast to steroid hormones, which do not differ structurally between the various vertebrate groups, gonadotropins differ between different classes, orders and even species of vertebrates. Alpha subunits are common and highly conserved among vertebrate classes, while β subunits are species specific (Yu 1997). This makes their quantitative measurement much more difficult than that of the steroid hormones.

133

Licht et al. (1979) first developed and applied homologous radioimmunoassays for protein hormones in a reptile, using the green sea turtle. When the original pituitary hormone purifications were done to construct the radioimmunoassays, a neurohypophyseal protein contaminant of about the same size as the FSH molecule was inadvertently included with the material injected into the rabbits for antibody production. This contamination dominated and invalidated the early assays for FSH, but new preparations eliminated this problem and made the assay highly specific (Licht and Papkoff 1985). The LH and FSH radioimmunoassays derived from green turtle pituitary hormones show a good cross reactivity with the gonadotropins of other chelonian species. Virtually all reliable quantitative data on circulating gonadotropins in chelonians are based on these assays. Like steroid hormones, levels of circulating gonadotropins are affected by capture, handling, confinement and captivity of chelonians, and show a rapid decline under these conditions (Licht et al. 1985a; Mendonça and Licht 1986a).

Several studies on freshly caught chelonians revealed the following relationships. Sea turtles show a marked periovulatory surge of LH in association with FSH and progesterone increases (Licht et al. 1979; Wibbels et al. 1992), but this periovulatory surge of LH has not been found in freshwater turtles in which LH remains more or less undetectable during the gonadal cycles of males and females (*Chrysemys picta*: Licht et al. 1985a; *Kinosternon odoratum*: Mendonça and Licht 1986a; *Chelydra serpentina*: Mahmoud and Licht 1997). LH was, however, positively identified in freshwater turtles: it increases, for example, in the plasma of *K. odoratum* after gonadectomy (Mendonça and Licht 1986a). In freshwater turtles, plasma LH may cycle at low concentrations below the detectable levels of the radioimmunoassays (RIAs).

In sea turtles, FSH tends to be slightly elevated during spermiogenesis and vitellogenesis and peaks together with LH in the periovulatory period (Wibbels et al. 1992; Owens 1997). In contrast, FSH levels are lowest in female *C. serpentina* during and after the time of ovulation, and rise progressively during vitellogenesis throughout the summer without being correlated with any of the gonadal steroids (Mahmoud and Licht 1997). In female *K. odoratum*, FSH remains undetectable throughout the year (Mendonça and Licht 1986a). In males of *C. picta* (Licht et al. 1985a) and *K. odoratum* (Mendonça and Licht 1986a), FSH increases seasonally concurrently with testosterone. In contrast, FSH rises at the onset of spermatocytogenesis in late spring in *C. serpentina* and plateaus for the rest of the summer without any specific peak, and without showing a close association with testosterone (Mahmoud and Licht 1997).

In conclusion, studies on circulating gonadotropins are still too limited, and the results too inconsistent, to resolve their role in the reproductive cycles of chelonians. In combination with in vitro studies, the results suggest that at least one function of LH may be to stimulate the production of progesterone by the turtle ovary, including the preovulatory follicles and corpora lutea (Licht 1982; Wibbels et al. 1992). Turtle LH and FSH have both been shown to stimulate progesterone production in ovaries (Crews and Licht 1975; Licht and Crews 1976) and testosterone formation in testes (Yu 1997) of turtles. The descriptive data for FSH also suggest a role in spermatogenesis and possibly male steroidogenesis.

7.3 Gonadotropin-Releasing Hormone

Neurosecretory neurones in the brain, mainly in the midbrain (hypothalamus), produce peptides which are released into the blood. These peptides function either as releasing hormones (RH) or inhibiting hormones (IH) that influence the release of hormone from a first order endocrine gland, for example the adenohypophysis. The release and secretion of gonadotropins is regulated by the gonadotropin-releasing hormone (GnRH), sometimes also called luteinising hormone-releasing hormone (LHRH). No direct studies on isolation and analysis have been published for a chelonian or reptilian GnRH.

Hypothalamic extracts of *Chrysemys picta* show seasonal variations in GnRH activity when tested in vitro on pituitary glands (Hall et al. 1978). Using a combination of chromatographic and radioimmunological techniques, King and Millar (1980) concluded that GnRH in the tortoise *Chersine angulata* was distinct from mammalian GnRH, but indistinguishable from that of birds and teleosts. During a comparative study of the distribution of brain peptides in the central nervous system of *Testudo hermanni* (Weindl and Kuchling 1982; Weindl et al. 1983a,b, 1984), no good or clear immunoreaction was observed by using mammalian LHRH antiserum, and neural pathways could not be visualised (A. Weindl and G. Kuchling, unpubl.). The GnRH of *T. hermanni* also seems to be immunochemically distinct from the mammalian molecule. A similar situation seems to occur in various other reptiles (Nozaki et al. 1984). Licht and Porter (1987), however, discussed unpublished data indicating that the brain of green sea turtles *Chelonia mydas* contains two distinct GnRH-like factors, one of which seems to be closely related to the mammalian GnRH and the other of which is not clearly related to any other GnRH identified so far; they also indicated that the hypothalamus contains about the same amount of immunoreactive GnRH as the remainder of the brain combined.

Licht and Porter (1987) reviewed in vivo and in vitro studies of GnRH regulation of gonadotropin release in turtles (*C. picta*, *Trachemys scripta* and *Kinosternon odoratum*) and concluded that in chelonians, as in amphibians and mammals, both gonadotropins (FSH and LH) respond similarly to a single GnRH, and that the dose sensitivity is not appreciably different from that of other tetrapods. Gonadal feedback seems to modify pituitary sensitivity to GnRH, but it is not known if this feedback involves direct effects at the pituitary as opposed to neural levels, or whether it involves receptors or gland content. Nothing is known about normal short-term temporal patterns of circulating GnRH or gonadotropins in chelonians. A key feature of the GnRH–pituitary axis in mammals is the pulsatile or episodic release of GnRH and the requirement for pulsatile stimulation at the pituitary level. Several discrepancies between in vitro and in vivo studies and among species in the responses of males and females still limit the understanding of how GnRH is integrated in the hypothalamus–pituitary–gonad axis of chelonians.

7.4 Pineal Hormones

The mechanisms which mediate the environmental information provided by photoperiod, temperature, rainfall and food supply (see Sect. 6.1) to the physiological systems regulating and modulating reproductive cycles are not entirely understood. It is, however, well known that the pineal organ or epiphysis of most vertebrates responds hormonally to changes in illumination. This organ produces several biological active substances including noradrenaline, serotonin and melatonin, the last of which is regarded as its true hormone. In some non-mammalian vertebrates, the pineal organ also responds electrophysiologically to illumination changes. In vertebrate phylogeny, the photoreceptor cells of the pineal organ undergo a gradual transformation from the photoreceptive cells in fishes and amphibians to the rudimentary photoreceptor cells in birds and to the secretory pinealocytes in mammals. The pineal system exhibits a high diversity in reptiles, extending from the eye-like parietal organ of some lizards to secretory pinealocytes in snakes. Chelonians have an intermediate position in this scale, and may develop either typical photoreceptors or secretory photoreceptors showing a modified outer segment or lacking one entirely (Meissl and Ueck 1980).

The pineal gland of *Trachemys scripta* is photoreceptive and functions as a light sensitive dosimeter. Sensory nerve cells in the pineal stalk show a spontaneous discharge during extracellular electrical recordings. This spontaneous discharge is inhibited by light stimuli of wavelengths between 400 and 750 nm, with a maximal sensitivity between 606 and 650 nm (Meissl and Ueck 1980). Very little is known about the role of the electrophysiological information originating in the pineal organ.

The daily pattern of melatonin secretion and release of vertebrates is entrained by the light–dark cycle. The principal sources of melatonin are the pineal gland and the eyes. Melatonin seems to be a characteristic product of ciliated photoreceptive cells. The daily pattern of melatonin secretion appears very similar in all vertebrates, melatonin concentrations always being much higher during the night than during the day in the pineal gland, the eyes as well as in the blood, whereas serotonin is synthesised and occurs in higher concentrations during the day. These patterns in the pineal and/or the blood were studied in four chelonian families: Testudinidae (*Testudo hermanni*: Vivien-Roels et al. 1979), Cheloniidae (*Chelonia mydas*: Owens et al. 1980), Trionychidae (*Lissemys punctata*: Mahapatra et al. 1986) and Emydidae (*Terrapene carolina*: Vivien-Roels et al. 1988).

In *T. hermanni*, both the maximum concentration and the amplitude of the daily fluctuations are increased during the breeding season in spring and early summer. The daily rhythm disappears completely during winter and hibernation (Vivien-Roels et al. 1979). Photoperiod appears to be an important factor in the regulation of the seasonal and daily variations of the pineal serotonin content, while temperature has a more drastic effect on the daily production of melatonin (Vivien-Roels and Arendt 1981). In *C. mydas*, mating and nesting adult females show the lowest serum melatonin levels (Owens et al. 1980) and a positive correlation exists between circulating melatonin levels and ambient temperature (Owens and Morris 1985). Serotonin is at its highest levels in *L. punctata* during the gonadal growth phase from April to June (Mahata-Mahapatra and Mahata 1991). In *T. carolina*, photoperiod influences the duration of the night rise of melatonin and environmental temperature influences the amplitude of the day–night rhythm (Vivien-Roels et al. 1988). Pineal serotonin levels of *L. punctata* decreased both at low (10 °C) and high (32 °C) ambient temperatures compared to controls at 25 °C (Mahapatra et al. 1989). Testosterone and oestrogen injected in juvenile male *L. punctata* cause a decrease of pineal serotonin which may be related to a stimulation of the synthesis of melatonin, while progesterone and corticosterone have the opposite effect (Mahata and Mahata 1992).

Owens (1980) suggested a functional link between reproduction and biological rhythms in sea turtles, and speculated that the pineal body could provide sea turtles with a system to translate the length of the day into an endocrine-based biological clock. Since more recent data show that both photoperiod and environmental temperature can modify the shape and amplitude of the pineal and circulating melatonin rhythm of chelonians, and since melatonin secretion may in turn be influenced by gonadal steroids, the pineal system of chelonians may not be restricted to the translation of photoperiodic information. It seems to be an important intermediary organ for the transformation of various environmental cues into hormonal signals which, in turn, may mediate long-term physiological changes such as seasonal reproduction and dormancy.

7.5 Other Hormones

Several hormones outside the hypothalamus–pituitary–gonad axis and the pineal system may also exert influence on reproductive events in vertebrates, for example thyroid hormones (thyroxine T_4 and triiodothyronine T_3) and prolactin, but studies on the function of these hormones in chelonians are very limited. Although Licht et al. (1985b) found no annual variation of T_4 in captive *Chelonia mydas*, Wibbels et al. (1986) detected a cycle in wild *Caretta caretta* and Owens (1997) discussed increased levels in *Lepidochelys kempii* correlating with follicular development and decreased levels during hibernation events of sea turtles. A reciprocal relationship between plasma androgen and T_4 occurs in wild male *Chrysemys picta*, with a peak in plasma T_4 coincident with the complete suppression of the spring rise in testosterone (Licht et al. 1985a). It is, however, unclear whether a reciprocal physiological relationship exists between the thyroid and gonadal cycles of chelonians or whether the separate sex steroid and thyroid cycles simply reflect independent or differential responsiveness of the gonads and thyroid to changing environmental stimuli. The latter is supported by in vitro studies of hatchling *Trachemys scripta* pituitary, testis and thyroid secretion, in which the temperature sensitivity of the thyroid and testes differed: thyroid glands were relatively insensitive to temperature and responded to thyrotropin between 12 and 32 °C, whereas testicular androgen secretion showed an abrupt decline in gonadotropin responsiveness below 28 °C (Licht et al. 1989).

The role of prolactin in chelonian reproduction is virtually unstudied, but prolactin is implicated in the control of reproductive activity and behaviour of many vertebrates, including reptiles (Mazzi and Vellano 1987). Prolactin also has a dehydration-preventing action in chelonians (Brewer and Ensor 1980a,b). The hypothalamic control of prolactin secretion in the adenohypophysis of *T. scripta* involves a stimulatory neurohormone, a control mechanism for prolactin which is more similar to that of birds – which have a hypothalamic prolactin-releasing hormone – than to that of mammals which involves both inhibitory and stimulatory neurohormones for prolactin (Fiorindo 1980). An endothelin 3-like substance is co-localised in the secretory granules which contain gonadotropins and thyrotropin in the pituitary of *Pelodiscus sinensis*. The expression of the endothelin 3-like substance is more pronounced during the breeding season (summer), more pronounced in males than in females, and stimulates in vitro the release of prolactin and growth hormone (Suzuki et al. 1997).

Nesting and oviposition have been studied in detail in sea turtles and several hormones seem to be involved in their regulation. A pronounced peak of arginine vasotocin (AVT) and neurophysin occurs during nesting, with both hormones reaching their peak in the circulation just as the first egg is dropped from the cloaca (Figler et al. 1989). Both AVT and neurophysin are synthesised in the neurohypophysis. AVT increases the contractility of the oviduct in *L. kempii* (Owens 1997). Prostaglandin F (PGF) and prostaglandin E_2 (PGE$_2$) also peak during oviposition. It is possible that AVT induces the synthesis of PGF in the oviduct, and that both hormones have roles in oviduct contraction. PGE$_2$ may have a role in cloacal/cervical relaxation (Guillette et al. 1991). All these hormones in concert facilitate contractility of the oviducts and rapid egg transport out of the oviducts into the nest.

Oxytocin – another neurohypophyseal peptide which differs from AVT by a single amino acid – contracts mammalian uterine muscles. In laboratory studies and captive breeding operations synthetic oxytocin is widely used in gravid chelonians to induce oviposition (Ewert 1985), but it is not well established whether oxytocin occurs as a natural hormone in chelonians. A study of neurohypophyseal peptides in the central nervous system of *Testudo hermanni* revealed, besides AVT and neurophysin, oxytocin-immunoreactive perikarya in the hypothalamus, with fibres projecting towards the neurohypophysis (Weindl et al. 1983b). The presence of an oxytocin-like peptide in the hypothalamo-hypophysial system was recently also demonstrated in *Mauremys leprosa* (Perezfigares et al. 1995, calling the taxon erroneously *Mauremys caspica*).

Corticosterone, an adrenal hormone which circulates in elevated levels as a response to stress (see Sect. 6.3), may be involved in the suppression of reproductive functions under stressful conditions. In addition, Whittier et al. (1997) found a high correlation of corticosterone and testosterone during the nesting season in female sea turtles (*C. caretta*), and hypothesised that corticosterone may play a concomitant role with testosterone in regulating ovarian and hepatic functions by mobilising lipid, carbohydrate and protein reserves during vitellogenesis. Growth hormone also induced hepatic cells cultured from male *C. picta* to produce vitellogenin in response to oestrogen treatment (Callard 1975; Ho et al. 1985), suggesting that a growth-hormone-like factor may act as a synergist to oestrogen in reptilian vitellogenesis (Ho 1987).

7.6 Hormones and Behaviour

The influence of hormones on behaviour of chelonians may be one of the least studied aspects of chelonian endocrinology, at least partly because attempts to find relationships between seasonal hormone levels and reproductive behaviour other than nesting largely failed. In addition, the behavioural trait of chelonians of retracting into their shells and not moving when disturbed somewhat discourages behavioural experiments.

"Turtles, for example, have long resisted the efforts of behavioural physiologists. We have not yet developed the ability to think like a turtle, to recognise what they are about, indeed to get them to do anything interesting. Yet this ability is critical for establishing what might be an experiment relevant to such an animal and what are the modes with which such an animal might react to the cues appropriately presented". (Crews and Gans 1992: 12).

The only manipulative experiments to relate male sexual behaviour of a turtle to circulating androgen levels are those by Mendonça (1987b) with *Kinosternon odoratum*, but the results are inconsistent and difficult to interpret. They demonstrate how captivity can disrupt reproductive physiology and behaviour of turtles (see Sect. 2.5) rather than how androgen levels can influence it. In their review of the physiological basis of sexual behaviour in male reptiles, Moore and Lindzey (1992) speculated that Mendonça's (1987b) experimental results are an expression of the intermediate reproductive pattern of temperate-zone turtles, which mate in autumn as well as in spring, and are intermediate between the associated and dissociated reproductive pattern of reptiles. Moore and Lindzey (1992) proposed that *K. odoratum* initiates courtship through either androgen-dependent or androgen-independent mechanisms, depending on the season, and that male sexual behaviours are expressed in the absence of androgen as long as the animals have been previously exposed to elevated levels of androgen (which is always the case when spermiogenesis has occurred). This seems to be an academic attempt to say that androgen levels may not be directly related to mating behaviour in male chelonians, an hypothesis supported by observations that many male chelonians potentially mate at any time during their activity period (see Sect. 4.2.3).

Indirect evidence suggests a regulatory role for sex steroids in the control of female sexual behaviour. The brain of *T. scripta* concentrates oestrogen in specific areas in the septal region, the preoptic area, the central hypothalamus, the amygdala and the hippocampal and piriform cortex (Kim et al. 1981). These oestrogen-concentrating areas are similar to those found in other reptiles, birds and mammals, suggesting similar endocrine control mechanisms that govern oestrogen regulation of reproductive behaviour in female chelonians (Whittier and Tokarz 1992). Several neuroendocrine peptides may have roles as central mediators of sexual behaviour, but this has not been investigated in chelonians. To summarise, we are still very far from understanding the physiological basis of the sexual behaviour of chelonians.

7.7 Conclusions

The neuroendocrine and endocrine systems perform the task of integrating the external and internal cues which regulate reproduction in such a way that gonadal products are produced and sexual behaviour is expressed at the appropriate time of the year. The pineal appears to be involved in the regulatory influence that photoperiod and temperature has on the chelonian reproductive system and may transform this information into humeral (e.g. melatonin) and nervous signals. The integration of other external cues is not well understood, and may be mediated through other sensory systems (e.g. eyes, olfactory, vomeronasal system). In the brain, specific nuclei which receive input from these sensory areas project to the hypothalamus where both internal and external cues appear to be integrated and where nervous commands are passed on to neurosecretory cells. These cells secrete releasing hormones or inhibiting hormones, for example the gonadotropin-releasing hormone that regulates release of gonadotropins from the adenohypophysis (anterior pituitary), which in turn stimulate gonadal maturation and steroid hormone production – the hypothalamus–pituitary–gonad axis. Steroid hormones feed back to the hypothalamus and the pituitary to modulate subsequent functions of these organs. A (for example, seasonal) change in the sensitivity of the gonads and/or other target organs to hormonal signals may also occur.

The notion that this cascade of neuroendocrine and endocrine factors, and the feedback loops described above, regulate all reproductive events is certainly overly simplistic. Several hormonal mechanisms outside the hypothalamus–pituitary–gonad axis are also involved in the regulation of reproductive processes. The neuroendocrine controlling mechanisms seem to be complex and may involve a number of different peptides which have been described in the central nervous system of chelonians, but the functions of which are still little known.

Chapter 8

Eggs and Embryonic Development

Chelonian eggs are buried in sand, soil or decaying vegetation, and typically away from standing water. Exceptions are the eggs of *Chelodina rugosa* (Kennett et al. 1993a) and *Dermatemys mawii* (Polisar 1996), which may be laid in nests under water (see Sect. 4.3.1), with the eggs in developmental arrest during the submerged period. However, in these species, too, the eggs eventually develop in slightly moist to dry soil or sand once the water recedes and the nests become dry. Since the eggs are typically left to develop without further attention from the parents (see Sect. 4.3.7), they have to be structured to protect the egg's contents from desiccation, bacteria, fungi and predation by soil arthropods. The shelled egg represents, in essence, the first environment that every turtle embryo is exposed to, and must provide all aspects of an embryo's needs from the time the eggs are independent of the female until the hatchling leaves the nest, a period that may exceed 1 year. In addition, the egg acts as container of material for preovulatory parental investment in care in the form of hatchling fat deposits and residual material in the yolk sac (Congdon and Gibbons 1990b) that provide the hatchling with energy for the often lengthy shift from the nest chamber to the habitat where it starts feeding and growing (see Sect. 4.4).

Egg characteristics, the embryology of chelonians and their sex determination mechanisms have been extensively reviewed in recent years (Ewert 1979, 1985; Miller 1985; Packard and Packard 1988; Congdon and Gibbons 1990b; Ewert and Nelson 1991; Ewert et al. 1994; Pieau 1996). In this chapter I will only provide a summary and an overview of the main topics and patterns. More detailed information can be found in those reviews.

8.1 Egg Characteristics

Once chelonian eggs are laid they remain underground or hidden under leaf litter and vegetation. All chelonian eggs are whitish with unpigmented, translucent eggshells that are commonly classified as pliable (or parchment-shelled), hard-expansible or brittle. Chelonian eggs are described as being spheres, ellipsoids or bicones. Spherical eggs are produced by large bodied species which often produce high numbers of eggs per clutch, in particular by the families Cheloniidae, Chelydridae, Trionychidae, Carettochelyidae and some large bodied species of the Testudinidae and Pelomedusidae. Kinosternidae and Emydidae and many members of other families produce some form of an elongated, symmetrical egg (Ewert 1979). Since morphological features of the females such as the diameter of the pelvic canal or the size of the posterior shell opening limit egg dimensions, elongated eggs can provide greater volume without increasing the short-side diameter of eggs beyond these constraints (Rose et al. 1996). Egg size is an important life history trait in chelonian species and populations and will be discussed in Section 9.1.2.

8.1.1 Eggshell

The chelonian eggshell (that is, all layers of a freshly laid egg external to the albumen) is composed of a mineralised, calcareous outer layer and fibrous inner layers. The texture of the shell reflects the degree of mineralisation. If the calcareous layer is thick or at least structurally cohesive, the eggshell is brittle and feels rigid. If the calcareous layer is thin or loosely cohesive, the fibrous membrane, which is always pliable, determines the texture. Chelonian eggs show a continuum from pliable to brittle, with hard-expansible eggshells in the middle. The shell membrane of flexible eggshells may have relatively thick, fibrous membranes, whereas rigid shells may have one or two thin shell membranes. The shell membranes consist of matted sheets or layers of fibres. The structural orientation of the fibres within individual sheets varies among groups and species from parallel to somewhat random. Their primary orientation changes direction from one sheet to the next (Ewert 1985).

The calcareous layer is formed of crystalline material, usually aragonite arranged into discrete shell units, but traces of calcite may form at the nucleation sites for shell units. Captive reared animals may produce eggshells with higher calcite portions, which indicates that characteristics of eggshells may be modified by the diet on which females are maintained. Although the ultrastructure of rigid eggshells appears at first very different from that of flexible eggshells, these differences may simply arise as a consequence of different amounts of time that crystallisation proceeds inside the oviducts of females (Packard and Packard 1988).

The type of eggshell determines the water relations of the egg with its environment. Packard et al. (1982), in reviewing the structures of eggshells and the water relations of reptilian eggs, distinguished three types: (1) thin, highly extensible eggshells with little or no calcareous layer (most lizards and snakes). These eggs are tightly coupled to the hydric environment; favourable hydric conditions are critical for a high hatching success and large hatchling size; (2) flexible eggshells with a well-defined calcareous layer (many chelonians). These eggs are relatively independent of the hydric environment, with hatching success not being unduly reduced by exposure of eggs to dry conditions, although hatchling size is reduced; (3) hard eggshells which are rigid and non-compliant (some chelonians, crocodiles, a few lizards) and therefore effectively independent of the hydric environment, for neither hatching success nor size of hatchlings is influenced by variations of the hydric environment of the eggs.

Mucous excretion accompanies egg laying in some chelonian species. The drying mucous forms a cuticle adhering to the outer surface of the calcareous layer, and seems to be an important barrier in some species, for example species of the Pelomedusidae: freshly laid eggs washed free of mucous dehydrate more easily, succumb to infection and appear generally less viable than coated eggs (Ewert 1985).

All chelonian eggshells tend to lose their integrity towards the end of incubation; the mineral layer may exfoliate as a granular powder. This change is due to loss of calcium from the inner surface of the mineral layer. The calcium probably is used by advanced embryos (Ewert 1985).

8.1.2 Egg Contents

In most freshly laid eggs, but in all unfertile eggs and in all eggs removed prematurely from oviducts, the vitelline sac is surrounded by albumen. The vitelline sac consists of the vitelline membrane (zona pelludica according to Guraya 1989) and its enclosed contents, the yolk and the embryo. The vitelline membrane arises in the ovarian follicle (see Sect. 3.2.3) and, together with the plasmalemma (cell membrane of the egg cell), separates the yolk from the albumen and maintains an osmotic gradient between them. According to the shape of the egg, the early vitelline sac is either elongate or spherical. In fertile eggs the embryonic disc, a gastrula at oviposition, is in the upper part of the vitelline sac. The adjacent vitelline membrane rises to the upper portion of the egg and adheres to the inner shell membrane. The eggshell at the point where adhesion has occurred becomes chalk-white against the translucent yellowish-white of the freshly laid egg. It first appears as a circular spot corresponding in size to the area opaca of the embryonic disk (Ewert 1979, 1985).

An overview of the composition of eggs regarding water content, dry mass and lipid content is provided by Congdon and Gibbons (1990b). Water averaged 68.8% of the wet mass of all studied chelonian eggs. Mean values of neutral (non-polar) egg yolk lipids range from 23.5 to 33.8%. Eggs of *Trachemys scripta* average 10.5 g in wet mass and 3.0 g in dry mass, with the shell dry mass averaging 18% of the dry mass of the egg. Polar lipids and non-polar lipids make up approximately 11 and 30% of the dry mass of the egg, respectively.

Soon after oviposition, the contents of the vitelline sac of fertile eggs become segregated into a lower opaque area that contains the viscous yellow yolk and an upper clear area that becomes the subgerminal space (Ewert 1985). This process may take place before the chalk-white spot appears (Knirr et al. 1997). Chelonian eggs have a large yolk at oviposition, but the yolk becomes even larger during the first 1–2 weeks of development and occupies almost the entire interior of viable eggs as water flows inward from the albumen. Thus, the yolk compartment at that time contains all of the lipid, all or most of the protein and water and part of the calcium for the growth and metabolism of the embryos (Packard and Packard 1988).

Air spaces in chelonian eggs reflect states of hydration. Pliable-shelled eggs respond rapidly to changes in moisture by expanding or indenting, whereas brittle-shelled eggs crack or form air spaces. During early to mid-development of brittle-shelled eggs, air pockets appear mainly interior to the fibrous shell membranes. In *Apalone spinifera*, the air space separates the inner and the outer shell membrane. During advanced development [at later stages than Yntema (1968), stage 20], there is an increasing tendency for air spaces to develop between the mineral layer and the fibrous shell membranes. In hard-expansible and pliable-shelled eggs, air spaces can also form in the albumen at any stage of development (Ewert 1985).

How do the brittle-shelled eggs of *Chelodina rugosa*, which are deposited in mud under shallow water, avoid water absorption and rupture? In that species, the period of submergence in water and developmental arrest can exceed 12 weeks without embryonic mortality. During the first week of submergence, the albumen loses considerable amounts of sodium through the shell and osmotic concentration drops from 234 mmol/kg at laying to 23 mmol/kg. The principal adaptation for the protracted developmental arrest under water is a vitelline membrane of such low permeability to water that the expansion of the yolk compartment occurs about 10 times more slowly than in other chelonians. Loss of embryonic viability is associated with contact of the vitelline membrane with the inside of the shell (Seymour et al. 1997).

8.2 Development and Hatching

8.2.1 Development Rates and Incubation Times

Chelonian eggs develop to late gastrulae in the oviducts, and further differentiation then ceases until eggs are laid (Ewert 1985). This pre-ovipositional developmental arrest causes all eggs to be oviposited in the late gastrula stage, independent of the time the eggs spend in the oviducts which may vary considerably (see Sect. 3.2.5).

Incubation period is the time which elapses from oviposition to pipping of the eggshell or hatching. Under natural or semi-natural conditions incubation times range from 28 days for *Pelodiscus sinensis* to 420 days for *Geochelone pardalis* (Ewert 1985). Among closely related species, those with more pliable eggshells tend, under favourable conditions, to have more rapid embryonic development and shorter incubation periods. For example, *Podocnemis expansa* has pliable-shelled eggs of 34–40 g which hatch at 42–47 days at 31–39 °C, the hard-expansible-shelled eggs of 15–31 g of *P. unifilis* take 75–90 days to hatch at 31–35 °C, and the brittle-shelled eggs of 13–20 g of *P. vogli* require 127–149 days at 30–41 °C (summarised from literature data by Ewert 1985). A similar relationship seems to exist in Malaysian batagurine turtles. Although some brittle-shelled eggs, such as those of the Trionychidae, develop rapidly, most brittle-shelled eggs are slow developing, and nearly all of the eggs taking over 150 days to hatch are brittle-shelled. Most pliable-shelled eggs develop rapidly or moderately so. Species that nest on sandbars and beaches lay rapidly developing and mostly pliable-shelled eggs, except for the rapidly developing brittle-shelled eggs of Trionychidae (Ewert 1979, 1985).

In the case of the three *Podocnemis* species mentioned above, the incubation period appears inversely correlated to egg size. This, however, is not a general trend. For chelonians overall, Ewert (1985) concluded that there is a weak trend favouring longer incubation periods for larger eggs. In particular, large eggs may be unable to develop as rapidly as small eggs when both are incubated at relatively cool temperatures. Large eggs may be at a disadvantage where growing seasons are both cool and short.

The incubation times of several species do not necessarily conform to the time spans needed for embryonic development, since periods of developmental arrest (other than the pre-ovipositional arrest) may affect incubation times. In *Chelodina rugosa*, which deposits eggs in underwater nests, it could be argued that the pre-ovipositional arrest is extended after oviposition, since the eggs only start further development once the nest dries out (Kennett et al. 1993b). Another form of developmental arrest of recently laid eggs is sustained by cool temperatures or "chilling", which can also be described as cold torpor. This occurs naturally during winter if eggs are laid in autumn, for example in the species *Kinosternon minor* and *Pseudemys floridana* (Ewert 1991). Some species, however, have post-ovipositional embryos that advance through early development extremely slowly even though the incubation temperature is conductive to rapid development of their later stages, as well as to rapid development of all stages in related species. This phenomenon is known as embryonic diapause and is found in the families Emydidae, Bataguridae, Kinosternidae and Chelidae. Some species require chilling to terminate the diapause, whereas others need elevated temperatures (Ewert 1991).

Embryonic aestivation, or delayed hatching, is a form of late embryonic dormancy that typically occurs under warm ambient conditions. It may last for weeks or several months and has been found in several species and families (Kinosternidae, Carettochelyidae, Emydidae, Bataguridae, Chelidae). Dry conditions tend to delay hatching in species with a capacity for embryonic aestivation, whereas dry conditions hasten hatching slightly in species which cannot aestivate, as in *Chrysemys picta* and *Chelydra serpentina* (Ewert 1985). A general assumption seems to be that embryonic aestivation is terminated by wet conditions, for example flooding of nests, and that a lack of oxygen is the cue for hatching (Köhler 1997). However, it is a drop in temperature which induces fully developed eggs of the Australian chelid *Pseudemydura umbrina* to hatch. Artificially incubated eggs of this species typically do not hatch at constant temperatures between 24 and 30 °C, but a drop of 3–4 °C induces hatching within a few hours or days (Burbidge et al. 1990). Eggs of the South American chelid *Phrynops geoffroanus* also do not hatch spontaneously, but need a wet substrate together with a drop of temperature to about 22 °C (Wicker 1997). Wicker speculates that water condensing on the eggs due to the temperature drop reduces the gas permeability of the eggshell, and that a lack of oxygen induces hatching in *P. geoffroanus*. This situation definitely does not occur in *P. umbrina*, the eggs of which also hatch under dry conditions as long as the temperature drops.

8.2.2 Descriptive Embryology

Descriptions of stages of embryogenesis of chelonians covering development from oviposition to hatching include the species *Chelydra serpentina* (Yntema 1968), *Chrysemys picta* (Mahmoud et al. 1973), *Lepidochelys olivacea* (Crastz 1982), Cheloniidae in general (Miller 1985) and *Testudo hermanni* (Guyot et al. 1994). A number of earlier studies which provide partial series of other species are discussed by Ewert (1985) who reviewed chelonian embryology, including organogenesis and developmental anatomy, excellently and exhaustively. Ewert correlates the development of organ systems with the 27 Yntema stages (0–26) and provides time-temperature relationships of developmental events and their correlation to the Yntema stages as determined by candling of living eggs of 37 chelonian species, an approach which does not harm the embryos and which is particularly useful to study the development of rare and endangered species. In this account I will only present and discuss some selected aspects of descriptive embryology that can be studied in live eggs by candling. Figures 8.1–8.6 give examples of the use of this technique in live eggs of *Pseudemydura umbrina*.

When candling eggs, it is useful to look for the extra-embryonic membranes, which are tissue outgrowths of the embryo itself, as they allow us to establish the earlier stages of development. These membranes include the yolk sac and allantois, which develop from endoderm and splanchnic mesoderm, and the amnion and chorion, which consist of ectoderm and somatic mesoderm. The following summary of the development of the extra-embryonic membranes is based on the account of Ewert (1985). The yolk sac membrane is the first extra-embryonic membrane to begin forming. The area pelludica, a blastular bilaminar layer of ectodermal and endodermal cells, represents the early yolk sac membrane. Later, as the mesoderm arises and spreads peripherally between the ectoderm and endoderm, the yolk sac membrane enters a trilaminar phase. The extraembryonic coelom then expands within the mesoderm, leading to a bilaminar yolk sac membrane of mesoderm plus endoderm. By Yntema stage 5, the mesoderm begins condensing into blood islands, and these define the area vasculosa of the vitelline circulation at stage 6. Rudimentary vitelline circulation is operative at stages 8–9 (Fig. 8.1). Candling allows us to follow the development of the blood vessels. The terminal sinus forms first (Fig. 8.2), and expands to cover greater areas of yolk (Fig. 8.3). Late in development, the vascular area comes to enclose the entire yolk mass. By this stage blood vessels also permeate throughout the yolk.

The amniochorion may function as a positioner for the embryo within the egg. It holds the embryo close to the eggshell enhancing respiration, but also holds the embryo a little away from the eggshell to prevent adhesion and provide space in which the dorsal region of the embryo can grow. The amniochorion first appears with the head fold in presomite embryos. At about Yntema stage 4 it forms a hood of ectoderm which then moves posteriorly over the embryo. The mesoderm spreads into the extraembryonic regions and forms a space within itself, the extraembryonic coelom, and gives rise to the amnion adjacent to the embryo, the chorion next to the vitelline membrane and the yolk sac membrane below. At

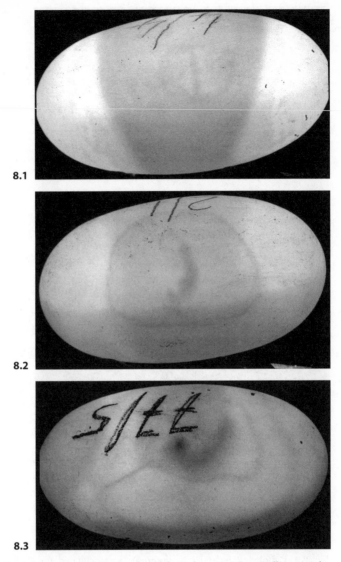

Fig. 8.1. Candled *Pseudemydura umbrina* egg, incubated for 8 days at 29 °C. Vitelline circulation (Yntema stages 8–9). During candling, the chalky-white band appears darker than the translucent parts of the eggshell

Fig. 8.2. Candled *Pseudemydura umbrina* egg, incubated for 20 days at 24 °C. The anterior vitelline veins are visible below the head (Yntema stages 10–11)

Fig. 8.3. Candled *Pseudemydura umbrina* egg, incubated for 26 days at 24 °C. Retina pigmented, vitelline circulation well developed (Yntema stages 12–13)

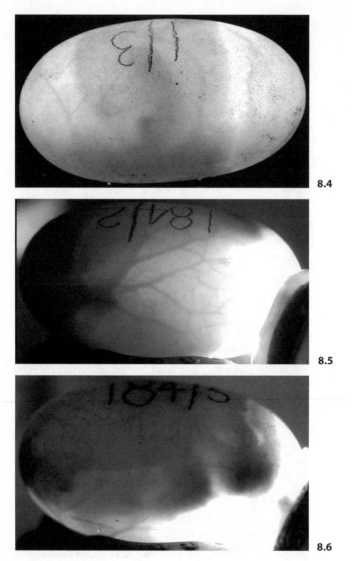

8.4

8.5

8.6

Fig. 8.4. Candled *Pseudemydura umbrina* egg, incubated for 32 days at 24 °C. The highly vascularised chorioallantois overlies the embryo (and amnion); its blood vessels are closer to the shell than those of the vitelline circulation and, therefore, much more clearly visible (Yntema stages 15–16)

Fig. 8.5. Candled *Pseudemydura umbrina* egg, incubated for 93 days at 24 °C. The chorioallantoic complex lines most of the shell membrane with clearly visible blood vessels; carapace and neck are lightly pigmented; embryos readily show movements, including thrashing with limbs (Yntema stage 20+)

Fig. 8.6. Candled *Pseudemydura umbrina* egg, incubated for 93 days at 29 °C. The allantois lines the entire shell membrane with blood vessels; near-term embryo, integumental pigmentation more or less completed; an air space between the mineral layer and the shell membrane is visible above the neck (Yntema stage 25)

about stages 21–22, the amnion is large enough to permit considerable movements of the embryo, including thrashing with outstretched limbs, without rupturing.

The allantois is the last extraembryonic membrane to form. It starts as a protrusion of the hindgut, and approaches and touches the chorion at about Yntema stage 13. The allantois then expands close to the shell membrane of the eggshell. It functions as the primary organ for mid- to advanced embryonic respiratory gas exchange, and as a receptacle for nitrogenous wastes. The highly vascular chorioallantoic complex then overlies the embryo and the amnion (Fig. 8.4). In advanced stages, the chorioallantois lines the entire shell membrane with blood vessels (Figs. 8.5, 8.6).

8.2.3 Pipping and Hatching

The fully developed embryo at Yntema stage 26 [or, according to the sea turtle staging system of Miller (1985), at stage 30] is ready to pip the eggshell. Just before pipping, the lobes of the allantois separate near the head and limbs. This parting and retraction of the allantois is probably initiated when the turtle tears and breaches the amnion. The allantois then gradually passes as an intact sheet over the top of the carapace and past the plastral bridges along its side. Most species pip while the allantois still covers the posterior part of the carapace. In kinosternids, however, the allantois usually completely withdraws before pipping to become a small pad adjacent to the umbilical opening (Ewert 1985).

The caruncle on the tip of the nose below the nostrils is often used to pip the eggshell, or the claws of the front feet pierce the eggshell. During pipping, most species have fairly large external yolk sacs. Testudinids and most emydids have large external yolk sacs at pipping and are not ready to leave the eggshell immediately. In these species, the time from pipping to hatching takes from several hours to days. Concurrent with the unfolding of the carapace and plastron which were compressed to fit the shape of the egg, the yolk sac withdraws into the hatchling coelom. The yolk may be forced into the hatchling by a shrinking of the yolk sac membrane. During the pipping stage of sea turtles, the yolk is partly withdrawn into the abdomen and is partially encased in a pigmented membrane (Miller 1985). When kinosternids pip, the yolk sac is already internalised in the embryo (Ewert 1979), and the smaller species may scarcely pause in completing emergence from their eggshells (Ewert 1985). In the chelid *Pseudemydura umbrina*, the external yolk is quickly internalised before or during pipping and hatching proceeds quickly. Typically it takes 5–10 min from pipping until the hatchling leaves the egg.

Video films of hatching *P. umbrina* eggs in the laboratory revealed that the first pip originates from a quick push of the head, with the caruncle cracking the eggshell, but typically not breaking a hole. A second push of the head may follow, but the shell is only broken by sideways pushes of the forelimbs some seconds later. Thrashing of the head and forelimbs than removes the end portion of the eggshell. Hatchlings leave the shell 5–10 min after pipping. At that stage, the yolk mass is totally withdrawn into the abdomen. The extra-embryonic membranes remain attached but appear pale and withered. When the hatchling leaves the remaining eggshell through the opening produced by the head and forelimbs, the extra-embryonic membranes tend to detach at the umbilicus and remain in the eggshell, or they detach shortly after the hatchling emerged from the egg. Natural nests do not provide the space available for hatching in the laboratory, and the whole process may be more protracted. *Pseudemydura umbrina* hatchlings dug out from wild nests before emergence were found with all eggshell fragments and the remains of the embryonic membranes neatly compacted under their plastron.

8.3 Sex Determination

In all classes of vertebrates, with the exception of reptiles, genotypic sex determination (GSD) is the rule, whereby the combination of maternal and paternal chromosome sets in the zygote determines the direction of development of the undifferentiated gonadal anlage towards male or female. In many chelonians, several lizards and, presumably, all crocodiles, the sex is determined phenotypically (or environmentally). In the latter system, no heteromorphic sex chromosomes have been identified and the sex is primarily determined by the temperature conditions during embryonic development, during the earlier stages of gonadal differentiation. In chelonians, temperature-dependent sex determination (TSD) is the most common of the environmental mechanisms of sex determination (Ewert and Nelson 1991; Ewert et al. 1994). The pioneering research into TSD in chelonians was done on *Testudo graeca* and *Emys orbicularis* (Pieau 1971, 1972).

Both Australian chelids studied so far have GSD (*Chelodina longicollis*: Georges 1988; *Emydura macquarii*: Thompson 1988), as well as the subfamily Staurotypinae of the Kinosternidae, the family Trionychidae, and *Clemmys insculpta* (Emydidae) (Ewert and Nelson 1991). Two patterns of TSD are known to occur in chelonians: one where males are produced at low temperatures and females at warm ones (Pattern Ia), and the other where cool temperatures produce females, intermediate temperatures males and warm temperatures females (Pattern II). These patterns tend to be associated with the direction of sexual dimorphism in adult size, especially for species with strong dimorphism. Pattern Ia occurs mainly in species in which adult females average larger than adult males; Pattern II occurs mainly in species with females smaller than males or in which body size is not dimorphic. The smaller gender typically arises at the coolest incubation temperature (Ewert and Nelson 1991). In the alligator snapping turtle *Macroclemys temminckii*, some embryos, in particular in clutches of small eggs, seem to be temperature-independent females because these individuals become females even at male-producing temperatures (Ewert et al. 1994).

An explanation of the adaptive value of TSD and sex ratio bias is the hypothesis of temperature-dependent differential fitness which predicts that the majority sex at each incubation temperature should have higher fitness (Charnov and Bull 1977). The research needed to test this hypothesis is logistically difficult, because laboratory fitness tests may not reflect the requirements in the field; but field tests are even more difficult, since the evolutionary significant parameter is lifetime reproductive fitness which is very hard to measure. Other hypotheses for the existence of TSD include phylogenetic inertia, but since GSD and TSD are both present in the most primitive chelonian families, it is difficult to determine the primitive state. Comparisons among phylogenetic sister groups suggest that, if TSD is primitive in chelonians, GSD has evolved independently at least four times; if GSD is primitive, TSD has evolved at least nine times (Burke 1993). Ewert et al. (1994) suggest that, of the two patterns of TSD in turtles, Pattern II may be basal or ancestral and Pattern Ia derived. Further adaptive explanation attempts for TSD include "group structured adaptation in sex ratios" and "sib-avoidance", which suggest that TSD may be either a by-product of selection for sex ratio biases in small breeding kin groups, or a mechanism for reducing inbreeding (Ewert and Nelson 1991). However, Burke (1993) found little support for the last two theories in a comparative analysis of chelonian species. In extending Charnov and Bull's hypothesis of temperature-dependent differential fitness, Roosenburg and Kelley (1996) proposed the "maternal condition-dependent choice hypothesis" for the maintenance of TSD. *Malaclemys terrapin* females which lay relatively large eggs were found to select open nesting sites with warmer conditions, whereas those females laying relatively smaller eggs place them in cooler sites. This is consistent with the prediction of the hypothesis, because egg mass is the primary determinant of hatchling mass and can result in as much as a 3-year difference in reaching minimum size of first reproduction in females, but may not affect age or size of first reproduction in males. In general, any demonstration of current selective advantage can only suggest why TSD persists, not why it evolved. Due to the long evolutionary history of chelonians and a continuously changing world, the reasons why TSD evolved may well have been dependent on conditions that no longer exist.

TSD is characterised by a pivotal temperature (sometimes also called "critical temperature" or "threshold temperature") that can be defined as the temperature during incubation which, if held constantly, gives 50% individuals of each sexual phenotype. The transitional range of temperature has two limits, that of maximum masculinisation and that of maximum feminisation. The thermosensitive period (or critical period) is the time (or stages of development) in which temperature changes affect the sexual differentiation. It generally corresponds to the first steps of histological differentiation of the gonads, somewhere between the end of the first and beginning of the second third of embryonic development (Mrosovsky and Pieau 1991). In most chelonian species, the thermosensitive period tends to lie somewhere between the Yntema stages 14 and 27 (Janzen and Paukstis 1991; Merchant-Larios et al. 1997). In *Lepidochelys olivacea*, the thermosensitive stages at male-promoting temperatures occur earlier (between stages 20 and 24) than at female-promoting temperatures (between stages 23 and

27) (Merchant-Larios et al. 1997). Pivotal temperatures are species- and population-specific and show geographic variations. Across central and southern North America, for example, they tend to increase with both latitude and longitude. These geographic trends probably reflect factors that affect nest temperatures (Ewert et al. 1994).

Most studies of sex determination have been conducted in the laboratory at constant temperatures. The thermal regime in natural field nests is more variable and complex than in the laboratory, with daily temperature fluctuations, seasonal differences and thermal gradients within single nests (Georges 1992; Souza and Vogt 1994; Valenzuela et al. 1997). In *Carettochelys insculpta*, the top and bottom eggs in single nests experience temperature differences of up to 3.5 °C. Warm nests produce only females and cool nests only males. In nests that produce both sexes, males emerge from the deepest and coolest eggs and females emerge from the shallowest eggs (Georges 1992). In natural nests of *Podocnemis unifilis*, the thermosensitive period is long (from Yntema stage 11 to 19), and the mean and variance in incubation temperature and the number of hours at or above the pivotal temperature are the indices that best describe sex ratios (Souza and Vogt 1994). Similar relationships were found in several other species (references in Valenzuela et al. 1997). In *Caretta caretta*, the sex of hatchlings depends on the daily proportion of embryonic development that occurs above the threshold temperature (Georges et al. 1994). Valenzuela et al. (1997), however, found that the parameters that best explain the sex ratios in natural nests of *Podocnemis expansa* are (1) the number of hours above 31 °C and (2) mean temperatures above 31 °C during day 29–30 of incubation. They call this short period that statistically best explains sex ratios the "thermoinfluential period". Georges (1992) also found that temperature experienced during only two incubation days (days 30 and 35 of incubation) best explained the sex ratios in field nests of *C. insculpta*. The biological significance of such short thermoinfluential periods is not known.

Sex determination is the event that sets sex differentiation in motion, a programmed cascade of events in which the indifferent gonad develops as an ovary or a testis with the appropriate urogenital structures. The effects of exogenous oestrogens, antioestrogens and aromatase inhibitors at different temperatures have demonstrated the involvement of oestrogens in the sexual differentiation of the chelonian gonads. Aromatase is the enzyme that converts androgens to oestrogens. Gonadal aromatase activity increases exponentially in differentiating ovaries, and remains low in differentiating testes. A thermosensitive factor may, directly or indirectly, intervene in the transcriptional regulation of the aromatase gene in reptiles with TSD (Pieau 1996). In addition, genes coding sex steroid hormone receptors may also be involved (Crews 1996). In *Trachemys scripta*, non-aromatisable androgens and their products seem to be involved in the initiation of male sex determination, whereas oestrogens and their aromatisable androgen precursors are involved in the initiation of female sex determination (Crews et al. 1996). After the period of temperature sensitivity, ovaries of *Emys orbicularis* embryos can still develop testis-like tubes or cords when aromatase activity – and thus oestrogen synthesis – is inhibited, showing that the ovary retains male potential also after the thermosensitive period (Dorizzi et al. 1996).

Merchant-Larios et al. (1997) suggested that, since irreversible molecular processes underlying sex determination occur earlier at male- than at female-promoting temperatures in *Lepidochelys olivacea*, the male sex may be the default state, and that the female condition must be imposed upon it. In contrast to placental mammals in which ovarian differentiation can occur in the absence of oestrogen or an oestrogen receptor, ovarian differentiation in chelonians (and other reptiles) requires the presence of oestrogen. In the absence of oestrogen, a testis develops (Lance 1997).

Sex steroid hormones appear to be the physiological equivalent of incubation temperature, and serve as the proximate trigger for male and female sex determination. Accordingly, sex steroids can also be used to establish the sex ratio of hatchlings. By collecting chorioallantoic/amniotic fluid of pipped eggs and by analysing metabolites of oestradiol-17ß and testosterone in this fluid, it is possible to use the oestradiol:testosterone ratio to sex hatchlings without the need to take blood samples. In *Caretta caretta* chorioallantoic/amniotic fluid this ratio is significantly lower in males (0.5 ± 0.1) than in females (2.2 ± 0.3) (Gross et al. 1995). Since traditional sexing techniques such as plasma sex steroid quantification or endoscopy are very difficult to use in hatchlings without damaging them, this is the only reliable, non-invasive and non-damaging method to sex hatchlings. Four to six hours after injection of 0.1 units of ovine FSH in hatchlings of *Podocnemis expansa*, testosterone was also measurable in blood samples of males, but not in those of females (Lance et al. 1992). However, this method seems to be more problematic since challenging hatchlings with gonadotropin may have unknown side effects. Furthermore, ovine FSH challenge did not produce measurable plasma testosterone levels in male hatchlings or 11-month-old diamond-back terrapins *Malaclemys terrapin* (Jeyasuria et al. 1994).

8.4 Effect of Incubation Environment on Neonates

As discussed with the adaptive hypotheses for the persistence of temperature-dependent sex determination in the previous section, the incubation environment may influence not only the sex of hatchlings, but also various fitness traits. The most important physical variables in chelonian nests are temperature, water and gas tensions, all of which show some interdependence on each other. One frequent product of variation in physical conditions in nests is variation in size of hatchlings. In most (but not all: see Janzen 1993a) turtles, cooler and wetter conditions produce larger chelonian hatchlings than warmer and drier conditions. Larger hatchlings may be more difficult for predators to capture and/or to handle and swallow than smaller conspecifics (Packard and Packard 1988).

In *Chelydra serpentina*, the hydric conditions during incubation also influence the locomotor performance of hatchlings. Due to a greater aerobic capacity, hatchlings from eggs incubated on relatively wet substrates were faster than those originating from a drier medium (Miller 1987). In the softshell turtle *Apalone mutica*, which shows temperature-independent sex determination, higher incubation temperature of eggs (30 °C) produces larger hatchlings that run and swim faster than hatchlings from eggs incubated at 28 or 26 °C (Janzen 1993a). In a field experiment, larger *C. serpentina* hatchlings exhibited significantly greater survivorship than smaller individuals during movements from the nest site to water. However, survivorship was not related to clutch, incubation condition (wet-dry, warm-cool) or locomotor performance; it was only significantly size dependent (Janzen 1993b). This result supports the simple "bigger is better" hypothesis for chelonian hatchlings.

On the other hand, larger hatchlings produced under wetter and cooler conditions have less residual yolk to sustain them after hatching than do smaller hatchlings from warmer, drier incubation conditions. Under some circumstances, a relatively large nutrient reserve may enhance survival more than large size. For example, hatchlings of several chelonian species may spend prolonged periods (winters, summers) in the nest cavity without feeding, during which time they are supported largely by lipids in the retracted yolk (Packard and Packard 1988).

An attempt to quantify fitness of hatchlings is to determine their early growth rate in the laboratory. The best studied turtle in this respect is *Chelydra serpentina*, but the methods applied as well as the results obtained in these studies are not uniform. Ryan et al. (1990) found that *C. serpentina* hatchlings from intermediate (26 °C) incubation temperatures had higher hatching masses, but turtles incubated at high temperature (30 °C) had higher rates of growth over the first 9 months. In contrast, Brooks et al. (1991) reported that *C. serpentina* hatchlings from eggs incubated at intermediate, male-producing temperatures (25.6 °C) grew to a larger size at 7 months post-hatching than those from low or high, female-producing temperatures (22.0 and 28.6 °C). Post-hatching growth rates in this study were independent of moisture effects. Hatchling mass (which was correlated to egg mass) was neither related to post-hatching survival nor rate of growth. Again studying *C. serpentina*, McKnight and Gutzke (1993) confirmed that intermediate incubation temperatures (27 °C) produce faster-growing hatchlings (monitored for 14 weeks) than low (22 °C) or high (31 °C) incubation temperatures, but found that 27 °C-hatchlings from dry incubation substrates grew significantly slower than those from wet substrates. They also found that maternal and/or paternal effects and social interactions influence growth rates. Bobyn and Brooks (1994) obtained similar results for *C. serpentina* with incubation temperatures of 22.0, 25.5 and 29.3 °C, with the intermediate temperature producing mainly males, with higher embryo and hatchling survival and post-hatching growth (monitored for 11 months), but substrate moisture (wet or dry) did not affect hatching success, post-hatching growth or post-hatching survival. Rhen and Lang (1995) dissociated experimentally the normally confounded effects of incubation temperature and sex by hormone manipulations of *C. serpentina* eggs, producing females at male temperatures and vice versa. The subsequent growth of hatchlings over 6 months was enhanced at normally male-producing incubation temperatures, but was not affected by sex.

In summary, in *C. serpentina* which expresses TSD Pattern II (see Sect. 8.3), hatchlings from intermediate, mainly male-producing temperatures, which have intermediate size and intermediate quantities of stored yolk, are generally the most "fit" in terms of growth and survival. In the desert tortoise *Gopherus agassizii*, which shows TSD Pattern Ia, male hatchlings produced at the upper end of the male-producing temperature range (30.6 °C) also grew significantly faster than female hatchlings produced at 32.8 °C, but also faster than males incubated at 28.1 °C (Spotila et al. 1994). Again, the intermediate, male-producing temperature produced the most "fit" hatchlings in terms of growth.

However, when *Graptemys kohni* eggs (TSD Pattern Ia) were incubated at 24, 27, 29 and 32 °C, embryonic survival was highest at 24 °C and lowest at 32 °C. Hatchlings from 32 °C-eggs (the primary female-producing temperature) were significantly smaller than those from other treatments, but they grew faster and 180 days post-hatching the turtles incubated at 32 °C were the same size as the turtles in the other treatments. In addition, survival to 1 year was greatest in turtles incubated at 32 °C and lowest in those at 24 °C (Etchberger et al. 1990). Hatchlings from *Malaclemys terrapin* eggs (also TSD Pattern Ia) incubated at 32 °C (exclusively females) also grew faster than those from eggs incubated at 26 °C (exclusively males). Egg mass still explained 59% of the variation in size among 3-year-old females, but not among 3-year-old males (Roosenburg and Kelley 1996). Therefore, in *G. kohni* as well as *M. terrapin*, hatchlings from high, female-producing temperatures which hatch with a small size and high quantities of stored yolk are the most "fit" in terms of growth and (at least in *G. kohni*) post-hatching survival.

In *G. agassizii* and *C. serpentina*, adult males reach larger sizes than adult females, whereas in *G. kohni* and *M. terrapin*, adult females are larger than adult males (Gibbons and Lovich 1990). It appears that, regardless of the relationship between sex and incubation temperature, the larger adult sex shows the fastest juvenile growth and the best survival under laboratory conditions. Under the (untested) assumption that post-hatching growth is correlated with lifetime reproductive fitness, the results of the studies above are in agreement with the Charnov-Bull model of temperature-dependent sex determination in reptiles, because they show a differential fitness effect of incubation temperature on the sexes.

Chapter 9

Reproductive Patterns and Life History

A life history can be defined as a suite of coevolved characteristics that directly influence population parameters (Congdon and Gibbons 1990a). Life history traits related to reproductive success of individuals include age and size at maturity, the proportion of assimilated energy devoted to reproduction and reproduction units (egg size, clutch size and clutch frequency), survivorship of offspring to reproductive age, adult survivorship and longevity. Because the contributions of fecundity and longevity to reproductive fitness are so direct, the importance of trade-offs is at once obvious. If there were no trade-offs, all organisms would be as fecund as the most fecund, as long-lived as the most long-lived, and would develop as rapidly as the most rapid developer (Smith 1991).

Chelonians are renowned for their longevity, delayed reproduction and iteroparity. These main life history characteristics seem to be related to the defensive hard shell which dominates chelonian life in interesting ways. The protective value of the armour increases as its owner grows. The construction of the shell demands resources, and selection seems to favour juveniles that invest in growth rather than reproduction. The costs of delayed maturity are offset by a potentially long reproductive life – individual chelonians may potentially reproduce over many decades. As a group, chelonians exhibit the greatest development of iteroparity and the lowest intrinsic rate of increase of any large order of tetrapods (Wilbur and Morin 1988).

Many aspects of the chelonian reproductive strategies discussed in previous chapters relate directly to life history. It is beyond the scope of this short monograph to discuss and review exhaustively the large body of literature on all aspects of chelonian life history. The life history evolution of chelonians was recently reviewed by Wilbur and Morin (1988), the life history and ecology of the North American slider turtle plus other (mainly American) turtles was the topic of a recent book (Gibbons 1990a), and the life history of sea turtles was reviewed by Chaloupka and Musick (1997). Since chelonian life histories have been extensively and excellently reviewed, I will only give a short overview and summary of the interrelationship of reproductive patterns and life history among chelonians, and focus disproportionally on recent discussions and findings, and on areas which have been omitted from all previous reviews, for example regeneration.

9.1 Resource Allocation

All organisms have to balance the allocation of resources to growth, maintenance and reproduction. Since natural selection operates through differences in reproductive fitness, energy invested in reproduction contributes directly to fitness, while the benefits from investment in soma are expected only in the future. Optimal resource allocation theories generally predict that it is optimal to grow early in life and, at a particular age or size, to stop growing and start reproducing at the maximum rate until the end of life (Koztowski 1992). Many chelonians, however, keep growing throughout much of their life, although growth slows once individuals reach sexual maturity: they show an indeterminant growth pattern (Andrews 1982). Chelonians, therefore, allocate resources simultaneously to growth and reproduction during much of their adult life. Indeterminant growth is often optimal when the death rate is a decreasing function of age (Engen and Sæther 1994), which seems to be the case in many chelonians: the shell's protective effectiveness increases with body size and the increased calcification that occurs with age. Since adult growth slows down with age, the proportion of energy invested in reproduction should increase as the individual grows older.

9.1.1 Growth and Maturity

Growth and growth rate is an important parameter of chelonian life histories. Some species, in particular those with pronounced annual periods of dormancy, display growth annuli on the epidermal scutes of the shell that can be used to estimate the age of an individual. In many other species, however, growth annuli are unreliable in determining ages of individuals. Skeletochronology, a method used to estimate age in a variety of vertebrates, requires the death of the turtle (Zug 1991). Recent results from two freshwater turtle species suggest that counting annual growth layers in claw sections can provide age estimates at least until maturity (Thomas et al. 1997). The best method to determine growth rates in wild populations is from the recapture of marked individuals. In various chelonians, the long-used von Bertalanffy growth model (non-linear regression) provides the best fit to population data (Andrews 1982; Kennett 1996; Lindeman 1997).

Seasonal temperature fluctuations in the temperate zone influence energy acquisition of chelonians and, therefore, their growth patterns. Two-year-old *Chrysemys picta* show a sigmoidal pattern of growth, with growth rates accelerating until late June and decelerating thereafter. This sigmoidal growth pattern does not strictly follow the seasonal variation in temperature (which directly affects the energy process rate); it also reflects the energy-allocation strategy relating to winter dormancy. Cooler temperatures in the early active season may inhibit activity and rate or efficiency of digestion, and feeding may not even commence for a period of weeks after emergence from dormancy. High temperatures late in summer may again inhibit activity, including feeding, as would declining temperatures in the autumn. Much of the energy harvested in autumn is probably allocated to storage for metabolic maintenance during hibernation, and depleted energy reserves may have to be partially replenished before growth is initiated again in the spring (Lindeman 1997). Adult turtles may have seasonal growth patterns different to those of juveniles, which can be largely explained by active-season temperatures and hibernation demands (plus resource availability), since the seasonal reproductive cycles of adults add a new dimension to the seasonal patterns of energy allocation.

The juvenile phase of chelonians lasts from several years to a few decades (e.g. in sea turtles; Miller 1997). The change from rapid juvenile growth to slow adult growth is pronounced. The body size at which the slope changes from the rapid growth of small animals to the slower growth of large animals can be estimated from recapture data by using a "split-stick" regression model (Kennett 1996). The change in slope is believed to occur at the onset of maturity (Bury 1979; Georges 1985; Kennett 1996).

Sexual maturity in female chelonians can be defined as the capability of producing eggs during the next breeding season (see Sect. 3.1). However, many researchers who studied sexual maturity in a turtle population did not consider that ovaries may start cycles of vitellogenesis some years before ovulation and egg production actually occurs. *Pseudemydura umbrina* ovaries, for example, start cycling at least 2 years before ovulation starts, with follicles reaching pre-ovulatory size and then being reabsorbed (see Sect. 3.1; Kuchling and Bradshaw 1993). This stage in the life of chelonians is best described as a subadult stage – the animals are neither juvenile nor truly mature or adult. This was often overlooked in turtle life history studies, possibly because of inadequate research techniques: dissection of vitellogenetic or pre-ovulatory females (Moll 1979) cannot give firm conclusions regarding ovulations and does not readily enable us to distinguish subadult from adult females. Dissection of females with oviductal eggs or corpora albicantia or palpation or X-ray techniques to monitor the occurrence of oviductal eggs (Gibbons 1990b; Gibbons and Green 1990) is appropriate to assess the onset of maturity and egg production, but does not provide information on the start of vitellogenetic cycles which may occur years before the first ovulation. The best methods to investigate the subadult stage in chelonians are non-lethal techniques, such as laparoscopy or ultra-sound scanning, which allow us to assess the ovarian conditions of individuals repeatedly (Chap. 2).

Since growth patterns relate to energy allocation, a significant change of the juvenile pattern of growth occurs with the start of vitellogenetic cycles when females reach the subadult stage. A second decrease of growth may occur when females reach maturity (start to produce eggs). Indirect evidence in subadult *Chelydra serpentina*, through the detection of a decrease in growth rate (a narrowing of growth zones on the carapace), suggests that vitellogenesis starts 2 or 3 years prior to first nesting in Iowa (Christiansen and Burken 1979) and, at the northern margin of their distribution in Algonquin Park, Ontario, about 6 years prior to first nesting (Galbraith et al. 1989). Female *C. serpentina* in Algonquin Park begin to divert energy from growth to the maturation of ovarian follicles between 11 and 15 years of age, which corresponds to mean carapace lengths between 197 and 236 mm. A second major decrease in growth rate (over 40%)

occurs between 17 and 19 years of age, at a carapace length between 249 and 258 mm, suggesting the beginning of oviposition at this stage (Galbraith et al. 1989).

These data on *C. serpentina* demonstrate that the subadult phase (vitellogenetic cycles without ovulation) is more protracted in northern, late-maturing populations (in which females reach reproductive maturity at a larger size and at a greater age) than in earlier-maturing southern populations. The same relationship may occur interspecifically in chelonians – that the subadult phase is longer when maturity is delayed. However, it is not always populations in cooler climates which mature at a larger size: *Trachemys scripta* mature at a larger size in some tropical areas than they do in temperate areas (Congdon and Gibbons 1990a). Chelonians, in general, do not seem to differ from other vertebrates and invertebrates in terms of the relationships between growth trajectories, ages at maturation and adult survival rates. Most turtles mature at aound 70% of their maximum size, and adult lifespans are proportional to age at maturity, relationships that are similar to those in other reptiles (Andrews 1982; Shine and Iverson 1995).

Growth rates, in particular of juveniles, show considerable variation among individuals in chelonian populations. Although the size of a growing individual must be correlated with its age (individuals become older as they grow larger), large individuals in a population are not necessarily the oldest ones. Variation in both juvenile growth rates and ages at maturity apparently causes most of the differences in body size of adults. Some of the ways that larger body size may increase a turtle's reproductive fitness are, for females, increased clutch size and increased size or quality of eggs and thus neonates, and for males, higher mating success and greater protection during extended excursions in search of females for mating. Indeterminant growth of individual chelonians may be a mechanism that enhances natural selection for longevity by increasing the proportion of late versus early births of individuals (Congdon and Gibbons 1990a).

9.1.2 Egg Size, Clutch Size and Clutch Frequency

A fundamental aspect of an organism's life history is the allocation of energy available for reproduction into units (eggs, clutches, reproductive bouts) that will simultaneously produce successful offspring and maximise the lifetime reproductive success of the female. Optimal egg size models assume that resources available for reproduction are limited and attempt to describe the relationships and interactions between egg size and number. They predict that most variation in reproductive output within a population due to environmental conditions will be in clutch size rather than egg size, because egg size has been optimised by natural selection (Congdon and Gibbons 1990b). However, in many chelonian species both egg size and clutch size increase with body size of females. In *Deirochelys reticularia*, approximately 75% of the increase in clutch mass associated with increased body size results from increased egg size. Egg size is positively related to body size of hatchlings and larger hatchlings are believed to have higher survivorship than smaller hatchlings. Increased egg size, thus, is a mechanism by which animals with indeterminant growth can increase the quality of their offspring that are produced later in life (Congdon and Gibbons 1990a). Chelonians as a group also show a significant negative relationship between latitude and egg size, after adjusting for body size (Iverson et al. 1993).

In chelonians generally, terrestrial species lay fewer and larger eggs for their size than freshwater species, and marine turtles lay the smallest eggs in the largest clutches. At both the generic and the family levels, egg mass and clutch size show positive correlations with adult carapace length, but after removing the effects of body size, there is a negative correlation between clutch size and egg mass. These comparative data support the idea of an evolutionary trade-off between clutch size and egg mass (Elgar and Heaphy 1989). After adjustment for body size, a significant, positive relationship exists between latitude and clutch size, but no relationship exists between latitude and clutch mass (Iverson et al. 1993). Interpopulation variations in egg size, clutch size and clutch mass in *Chrysemys picta bellii* also reveal an egg size/clutch size trade-off at the intraspecific level (Rowe 1994).

Reproductive output not only depends on clutch size, because many chelonian populations produce several clutches each year while others do not even reproduce annually. Multiple clutches generally reflect the morphological constraint of packing shelled eggs within the body, rather than energy acquisition during the nesting period, since most of the energy (yolk) accumulation typically takes place over a long time span prior to the nesting season (Chap. 5). Iverson (1992b) examined the effects of various parameters on individual as well as total annual clutch mass (clutch mass/clutch frequency) of chelonians. Body size-adjusted individual clutch mass is strongly, positively correlated with latitude and weakly with diet (higher in carnivorous species). Habitat (i.e. marine, freshwater or terrestrial) has an effect on neither size-adjusted clutch mass nor size-adjusted annual clutch mass. Body size-adjusted total annual clutch mass shows a strong, negative correlation with average age at maturity: it is higher in early-maturing species. Since nesting seasons at higher latitudes are shorter, species at high latitudes have lower clutch frequencies, but higher size-adjusted individual clutch mass. However, latitude has no effect on size-adjusted annual clutch mass or its negative correlate, age at maturity.

Because age at maturity is generally positively correlated with survivorship and longevity (an assumption which is still untested in chelonians), variation in the latter may be the primary evolutionary determinant of variation in annual reproductive output of chelonians. This would conform to the hypothesis that high mortality schedules select for early maturity (and hence increase reproductive effort per bout) and that, therefore, reproductive effort is inversely correlated with longevity (Williams 1966, 1992). However, populations of turtles in which the rate of resource acquisition is slow would also be expected to have delayed maturity because their growth would be slower. Because of the slow rate of resource acquisition, less energy should be available per annum for reproducing adults (Iverson 1992b). Local environmental conditions (proximate factors) often dictate the specifics of reproductive output through their effects on rates of resource acquisition (see Sect. 6.1.3). The relative influence of proximate environmental versus ultimate genetic control over annual reproductive output in chelonians is still largely unexplored.

9.1.3 Reproductive Effort and Cost of Reproduction

A characteristic of most models which try to explain the evolution of life histories is the prediction of an optimum reproductive effort for a certain breeding season or a certain age of an organism. With an increase in reproductive effort, most models predict a decline of the residual reproductive value (future reproductive output), a phenomenon commonly known as the "cost of reproduction". This trade-off between the present and the future reproductive output determines which will be the optimum reproductive effort, which maximises the sum of the present and the future reproductive output (Bailey 1992).

The physiological model of how materials and energy are acquired, processed and utilised is based on rates – feeding rates, metabolic rates and excretions rate, growth rates and reproductive rates. The energy used per unit of time is the critical measurement. The ratio of reproductive biomass to the mass of the parent, or of calories expended for reproduction to calories of the parent, are static measures that do not represent the proportion of energy flowing through the organism that is devoted to reproduction (Stearns 1992). If these criteria are applied to chelonian papers, few actually measured reproductive effort; what is commonly measured is the quantity of reproduction. From the point of view of demographic theory, to measure the quantity of reproduction and the cost of reproduction is, however, adequate (Stearns 1992).

Chelonian populations and individuals often show high variations in clutch size and frequency between years, a phenomenon supporting the idea that annual reproductive output is largely governed by extrinsic environmental factors (Gibbons and Green 1990). Environmental variability, thus, will confound and mask heritable patterns of reproduction. If resource availability and acquisition by the individual control reproductive output, as often seems to be the case (see Sect. 6.1.3), then reproductive output would be expected to show particularly high annual variability in unpredictable environments. Supportive evidence for this hypothesis can be found in Section 5.4. Costs of reproduction should be more evident in capital breeders (e.g. the green sea turtle *Chelonia mydas*) which use stored energy for reproduction than in income breeders (e.g. the Aldabra tortoise *Geochelone gigantea*: see Sect. 6.1.3), in which the direct costing relates more directly to behavioural decisions, such as where or how long to forage each day. Income and capital breeders are the ends of a continuum; most species mix the two modes. For example, *Pseudemydura umbrina* can be classified as a "capital" breeder regarding vitellogenesis, since most of the energy is allocated to the follicles when the females are dormant and do not feed, but regarding ovulation the species is an "income" breeder which requires the immediate harvest of a large amount of resources in order to ovulate and, actually, produce eggs.

9.1.4 Mortality, Longevity and Ageing

Although data on the relative survivorship of adult and juvenile chelonians are scarce, mortality rates of eggs, embryos in the nest, first-year hatchlings and juveniles are apparently much higher and more variable than those of adults. For individuals, shell strength increases with body size and the increased calcification that occurs with age and, therefore, functions more effectively in the protection of adults than of hatchlings and juveniles. Frazer et al. (1990) reviewed and discussed survival rates for different age groups of various species and concluded that, although turtles are among the longest-lived vertebrates, most individuals die during their first or second year of life. The majority of nesting studies indicate a clear pattern of low egg and hatchling survivorship in a wide variety of species and families. However, for adults and large juveniles, annual survival rates of greater than 90% are not unusual in unexploited chelonian populations inhabiting relatively stable habitats. Turtle populations that experience relatively high adult survival rates also tend to exhibit delayed maturation (Shine and Iverson 1995).

Reproduction in itself involves physiological as well as other costs, for example increased risk of predation or desiccation during nesting, and there is a trade-off between current reproduction and survival. The various mechanisms which allow chelonians to reproduce in unpredictable habitats (see Sect. 5.4) seem to have the primary function of modifying reproduction in order to minimise mortality of the parent. A mechanism that enhances natural selection for longevity by increasing the proportion of late versus early births of an individual may be indeterminant growth, operating through low mortality, increased survivorship and/or increased reproduction associated with large body size.

The dichotomy between species that mature early, have many offspring, make a large reproductive effort and die young, and species that mature later, have few offspring, make a small reproductive effort and live a long time is often referred to, respectively, as "r-selected" and "K-selected" (Pianka 1970). This hypothesis assumes that r-selected species evolved under density-independent conditions which selected for a high intrinsic rate of increase (r), whereas K-selected species evolved under density-dependent conditions that selected for ability to withstand high densities of conspecifics (K represents saturation density). This deterministic concept was recently rejected in favour of stochastic models that analyse the age-specific effects of mortality as well as the mode of population regulation that generates the mortality (Stearns 1992). Selection pressure for a longer reproductive lifespan can be generated by variation in offspring survival rates from one reproductive attempt to the next. The environment varies in space and time with respect to offspring chances. The more variable the environment, the more samples (clutches) must be taken by the parent to achieve a given probability that a certain mean number of offspring survive to reproduce. This process is called "bet-hedging" after the practice of professional gamblers: the best way to produce a steady long-term profit is to spread the money over several bets (Stearns 1992). All these theories that attempt to predict associations of life history characteristics based on single selective factors have, overall, not been very useful in chelonian studies, in which more specific theories that attempt to predict the direction of selection on single traits seem to hold more promise (Congdon and Gibbons 1990a). The reason for this may be that chelonians as a group are already clustered at one far end of the range of the phenomena that these concepts try to explain. However, Cunnington and Brooks (1996) compared two snapping turtle (*Chelydra serpentina*) populations with loggerhead sea turtles (*Caretta caretta*) and concluded that in *C. serpentina*, but not in *C. caretta*, increased survival in the first year can yield a positive population growth rate. Turtle eggs and hatchlings produced in "bumper" years can maintain a stable population in *C. serpentina* which, thus, shows a high degree of bet-hedging.

Because chelonians are among the longest-lived vertebrates, concepts relating to senescence are of obvious interest. Despite some earlier speculations by various researchers that reproductive senility may occur in old females, a re-evaluation of data led Congdon and Gibbons (1990a) to conclude that there are no indications for reproductive senility in chelonians. The same assumption proved to be of critical importance for the survival of *Pseudemydura umbrina*, the world's rarest turtle: when the last captive females of *P. umbrina*, which were collected as adults in the 1950s, stopped breeding in the 1980s, wildlife managers thought that reproductive senility was a likely cause. However, direct investigation of the ovarian activity by ultra-sound scanning disproved this concept, and changes in captive management re-established the reproductive activity of all these old females (Kuchling 1988a; Kuchling and DeJose 1989; Kuchling and Bradshaw 1993) which, currently, still reproduce successfully.

It is difficult to collect reliable evidence for senility in wild chelonians since, even with high adult survivorship, few very old individuals will exist in a population. Since chelonians do not affect the fitness of their offspring once nesting is completed (e.g. through parental care), natural selection should favour postponement of senescence in an individual only during its reproductive life, and reproductive failure and death should be closely linked in time. Congdon and Gibbons (1990a) suggest that the protective shell allowed chelonians to evolve longevity by (1) decreasing adult deaths due to extrinsic causes, (2) increasing the proportion of late births relative to early births, (3) allowing more adults to live long enough for senescence traits to become implicated in their deaths, and (4) permitting natural selection to act on age-of-onset modifiers, and suppression genes to act on traits of senescence. This concept is consistent with the antagonistic pleiotropy theory of senescence (Williams 1957), an adaptive explanation of ageing which requires genes that have opposite effects on fitness when expressed at early and late ages. A variant of the theory of antagonistic pleiotropy is the disposable soma theory, in which ageing results from a low level of repair of somatic damage, whose accumulation during life leads to a deterioration of an organism's physiology (Kirkwood 1981).

9.1.5 Somatic Repair

The relationship between repair and reproduction is central to both short-term and long-term survival of animals. In the short term, extra investment in repair

164

increases expectation of life but decreases the resources available for reproduction. In the long term, genotypes that optimise the balance between repair and reproduction will be more likely to survive. The disposable soma theory predicts that longer-lived species should have better somatic repair mechanisms and regeneration capabilities (Kirkwood 1981). Using a dynamic programming model which considers the whole lifetime of an organism together with the three resource-demanding activities growth, reproduction and repair, Cichon (1997) found that a low extrinsic (environmentally caused) mortality rate and high repair efficiency promote allocation of resources to repair, resulting in low growth rates, delayed maturity, delayed ageing and dramatic enhancement of survival and maximum lifespan. This model also suggests that selection for late reproduction may favour investment in repair, delayed ageing, lengthened lifespan, delayed maturity and increased body size. Since chelonians are extremely long-lived and exhibit the greatest development of iteroparity and the lowest intrinsic rate of increase of any large order of tetrapods (Wilbur and Morin 1988), they are excellent candidates to test hypotheses of the evolution of repair mechanisms and their relationship to other life history traits.

There are two main types of somatic repair or regeneration: tissue regeneration, a basic and simple repair mechanism, is the proliferation of cells in a specific tissue through mitotic divisions; epimorphic regeneration involves epidermal wound healing, dedifferentiation of cells and redifferentiation into several tissue types (Goss 1969). Structures of little importance should not be under strong selection pressure for regeneration. Conversely, the loss of a structure which is essential for immediate survival will not allow the organism to live long enough to provide selection pressure for the regeneration of that structure. Thus, structures of intermediate value – not essential for immediate survival but important enough to warrant their redevelopment to regain selective advantages – should be most likely to evolve regenerative abilities (Goss 1969; Reichman 1984). The chelonian shell seems to possess these attributes.

Although, at first glance, the shell may seem essential for immediate survival, chelonians can survive extensive carapace damage as long as the remains of the old bony shell serve as temporary protection while a new carapace is produced. The chelonian carapace shows a considerable repair and regenerative capacity after damage and large scale destruction (Bellairs and Bryant 1985). Here I will explore in a comparative approach the epimorphic regeneration of the chelonian carapace after damage through traumatic injury, fire or predation events, but limit the discussion of carapace regeneration largely to its relationship with survival, indeterminant growth and reproduction. Descriptions of the actual process of carapace regeneration are provided by Gadow (1886), Smith (1958), Rose (1986) and Kuchling (1997a).

Populations of terrestrial chelonians sometimes show a high prevalence of shell damage. Apart from mechanical damage, e.g. by hoofed animals, predators, machinery or cars, a frequent cause is fire. For example, a severe summer fire in the habitat of *Testudo hermanni* in northern Greece caused an estimated 40% population reduction, with juveniles showing the highest mortality, while large adults survived best; many of those remained in shallow depressions (pallets) and were scarred when the fire swept through rapidly, showing burn marks to the rear of the carapace (Stubbs et al. 1985). Such burn damage may destroy virtually the whole epidermal layer of the carapace, and affect and expose most of the carapacial bone plates without killing the animal (Kuchling 1997a). Many terrestrial chelonians in mediterranean and savanna climates may experience fires during their lifetimes, and the need for shell regeneration is common in terrestrial species. In contrast to most mechanical injuries, fire damage often affects relatively large areas of the carapace. During large scale regeneration of the carapace a new shell is formed under the damaged bone layers which are eventually cast off (Fig. 9.1). The regenerated carapace, owing to the fact that it is formed under the old damaged one, is slightly smaller than the old armour which it replaces.

When the chelonian shell increases in size, growth of the dermal bone plates takes place at their sutures. The epidermal scutes grow by gradual deposition of new horny material over their entire undersurface by a layer of Malpighian cells which propagate laterally along the scute seams. This leads to concentric growth lines of horn scutes. Growth of individual scutes typically is not the same in all directions and, depending on scute and species, may even be extremely asymmetrical (Zangerl 1969). The subdivision of the horny surface cover of the shell into a mosaic of scutes facilitates its proper growth according to its species-

Fig. 9.1. Adult female *Testudo hermanni*. Regenerating carapace after fire damage; dead bone plates of the old carapace have already been partly cast off in the central and posterior area of the carapace and have lifted off and are about to be shed in the right posterior costal area (Kuchling 1997a)

specific shape. Until recently, most large scale regenerations were observed in adult box turtles of the genus *Terrapene* (Smith 1958; Legler 1960; Rose 1986) in which the regenerated shell shows no evidence of the typical horn scute pattern (Fig. 9.2). This led to the general conclusion that adult chelonians are not able to reconstitute the typical horn scute pattern during carapace regeneration (Zangerl 1969; Bellairs and Bryant 1985). However, a regenerated shell without scute seams may have a very limited or no potential for further growth.

Growth of most chelonians is indeterminate (see Sect. 9.1.1) and they spread their reproductive effort over most of their adult lifespan (see Sect. 9.1.3). Theoretically the turtle shell keeps growing throughout life, although in very old chelonians the bone sutures may ossify completely and growth ceases. American box turtles (*Terrapene*) are exceptional in so far as their bone sutures ossify and the shell stops growing while they are quite young adults (Pritchard 1979a). A box-like armour sets relatively rigid limits to the quantities of food material, water, air, energy stores, waste products and, in the case of females, eggs which the body can accommodate. As discussed above (see Sects. 9.1.1 and 9.1.2), reproductive traits in several chelonian species indicate that large (and often old) females have a higher reproductive success and higher quality offspring than smaller ones. This suggests that, for most chelonians, continued growth well into adulthood may be an important trait which increases their reproductive fitness.

Fig. 9.2. Adult male *Terrapene ornata*. Upper portion of carapace regenerated without scute seams (Kuchling 1997a)

Since natural selection operates through differences in reproductive fitness, any repair mechanisms should evolve to maximise future reproductive success of the individual. For many chelonians this may include the potential for further shell growth after carapace regeneration. Tortoises of the family Testudinidae (suborder Cryptodira) are, indeed, capable of restoring the corneal scute pattern during regeneration (Fig. 9.3) and, therefore, retain the capacity for further growth. Figure 9.4 shows substantial growth of a radiated tortoise shell after a regeneration event of the carapace, as indicated by the growth rings along the margins of the regenerated vertebral scutes. The capacity to regenerate the epidermal scute mosaic also occurs in the semi-aquatic *Geoemyda yuwonoi* (Bataguridae, suborder Cryptodira), as demonstrated by a photograph of Fritz and Obst (1996), and in the freshwater turtle *Pelusios subniger* (Pelomedusidae, suborder Pleurodira; Kuchling 1997a). Pleurodire and cryptodire chelonians have represented separate lineages since an early stage of turtle evolution (Gaffney et al. 1991). The regenerative capacity of restoring the typical epidermal shield pattern seems to be a primitive feature of chelonians, which may have been lost in the box turtles of the genus *Terrapene* (family Emydidae) which have reached a relatively high degree of evolutionary specialisation.

From the perspective of natural selection, repair is only beneficial to the extent that it improves the individual's future reproductive output and fitness. There should be a limit beyond which it is not worthwhile to invest in reconstruction of particular details of the regenerated organ. A balance must also be struck between somatic longevity and reproductive capacity, taking into account the rate of extrinsic (environmentally induced) mortality. The box turtle *Terrapene ornata*, for example, has a predilection for dung insects and for moving along ungulate pathways, which exposes them to the risk of being trampled and damaged by hoofed animals. Most adults of wild populations studied were found to have one or more injuries on the carapace that have healed or were undergoing repair (Legler 1960). It may be a general trait in the genus *Terrapene* to trade increased shell strength against the potential for future (and indeterminant) growth relatively early during adulthood through ankylosis, the complete ossification of shell sutures and fusion of bone plates, whereupon growth ceases (Pritchard 1979a). In adult box turtles in general, as well as in those individuals which regenerate their carapace, heavy ossification and strength of the armour may be critically important for survival and longevity in close quarters with hoofed animals and, therefore, for the future reproductive output of the individual – it has to stay alive in order to reproduce. In balance, the potential for further growth may be negligible for reproductive fitness in adult box turtles and would impair the quick restoration of a strong shell.

In this regard it is also interesting to note that, in *Terrapene ornata luteola* for example, egg size is not correlated to female body size (Nieuwolt-Dacanay 1997), whereas such a correlation exists in many other chelonians (see Sect. 9.1.2). This suggests that the trait of increasing the quality of the offspring that are produced late in life is not of particular importance to *Terrapene*. In chelonians with different life styles and indeterminant growth, the resumption of growth after carapace regeneration, which necessitates the restoration of epidermal shield seams, may

Fig. 9.3. Adult female *Testudo graeca ibera*. Near total regeneration of carapace with scute seams. The seams defining the anterior four vertebral scutes are more clearly developed than those of posterior and lateral portions of the carapace where additional small scutes and tubercles can be seen along the seams of the typical pattern

Fig. 9.4. Adult female *Geochelone radiata* with regenerated second and third vertebral scute; note growth rings along margins of regenerated scutes (*arrow*)

be more significant for their future reproductive success. In all European tortoises (*Testudo hermanni, T. graeca, T. marginata*) in which adults readily regenerate scute seams (Kuchling 1997a), egg size is positively correlated to female body size (Hailey and Loumbourdis 1988, 1990).

Older individuals of any organism show reduced regenerative abilities (Goss 1969), and larger individuals are less likely to regenerate successfully than smaller individuals (Reichman 1984). A loss of scute seams during regeneration and, presumably, cessation of further shell growth may not significantly reduce the reproductive fitness of adult box turtles, but in small, juvenile box turtles, the resumption of growth after carapace regeneration should be of selective advantage. Young and small box turtles should still have the capacity to restore at least some scute seams during carapace regeneration. This, however, remains to be investigated (Kuchling 1997a).

Due to their long evolutionary history, diversity of life styles and ranges of adult sizes, growth patterns and reproductive strategies, chelonians offer excellent models to study the costs and benefits of somatic repair and to test evolutionary theories. Their impressive repair efficiency and the variability of their regenerative capability support the hypothesis that a strong relationship exists between repair and the life history traits of low extrinsic mortality, indeterminate growth, longevity and delayed maturity.

9.2 Population Structure and Dynamics

A population can be regarded as a set of individuals with a common life cycle, which is a sequence of morphological states structured by age and/or development stages. Population dynamics concerns temporal changes in abundance and spatial structure. In addition to the aspects discussed under resource allocation (see Sect. 9.1), important parameters to understand population structure and dynamics include recruitment, dispersal, density, activity range or home range, emigration, immigration, age distribution and sex ratio. All this information is needed in order to develop population models. Since, worldwide, most chelonians are in decline, an understanding of population dynamics is essential to evaluate the health and viability of populations and to decide what risk management strategies might be appropriate for the protection of populations or species. A limiting factor for our understanding of chelonian population dynamics is the need for long-term studies over many years or decades in order to assess variability of measured characteristics due to good and bad seasons or years.

Population structures and dynamics have been reviewed for marine turtles (Bustard 1979; Chaloupka and Musick 1997), terrestrial chelonians (Auffenberg and Iverson 1979) and freshwater turtles (Bury 1979). A number of recent papers provide life tables or equivalent data for freshwater turtle populations, including populations of *Chrysemys picta* (Wilbur 1975; Tinkle et al. 1981), *Trachemys scripta* (Frazer et al. 1990), *Kinosternon flavescens* (Iverson 1991), *Kinosternon subrubrum* (Frazer et al. 1991), *Emydoidea blandingii* (Congdon et al. 1993) and *Chelydra serpentina* (Congdon et al. 1994; Cunnington and Brooks 1996). Relatively good data, including population analyses by matrix models, are available for sea turtle species, and were reviewed by Chalupka and Musick (1997). Good data sets for terrestrial chelonians are scarce, but a population analysis by matrix model is available for the desert tortoise *Gopherus agassizii* (Doak et al. 1994).

Chalupka and Musick (1997) provide an overview and a critical analysis of the major methodological issues concerning population modelling of sea turtles, their discussion being relevant for all chelonians. A main purpose of population modelling is to explore how a population responds to natural and human-induced perturbations. The abundance of individuals over time in each state is determined by five basic demographic processes or rates (birth, death, growth, immigration, emigration). A fundamental limitation of the various population modelling exercises is that many inputs of important parameters, for example on size- and age-specific mortality rates, are often guesses or only supported by weak data. The main value of numerical models is, therefore, heuristic – that is, to challenge mind sets and assumptions and to encourage more focused empirical research.

What the models can and do show for the populations of the long-lived chelonians is that population stability is most sensitive to changes in adult, subadult and larger juvenile survival and much less sensitive to changes in fecundity, nest survival and age at sexual maturity. It is mainly the high rate of adult survival, and the long life it confers, that allows chelonian populations to tolerate low and variable egg and first-year juvenile survival (see Sect. 9.1.4). This has important implications for conservation programmes. When subadult and adult survival is reduced, protection of nest sites and/or headstarting to increase survival of small juveniles cannot, in the absence of programmes to reduce mortality of older juveniles, subadults and adults, compensate for losses in those later stages (Congdon et al. 1993; Heppell et al. 1996). Headstarting, however, can augment chelonian populations when adult survival is near its maximum value (Cunnington and Brooks 1996) or, in recovering populations, after management actions have returned adult survival to high levels (Heppell et al. 1996).

Chapter 10

Threats, Reproduction and Conservation

The major threats to the persistence of chelonian species around the world are over-exploitation by humans, habitat loss and environmental degradation. That exploitation for food can extirpate chelonians was already convincingly demonstrated as early as the eighteenth century, when six out of eight species of giant tortoises on Indian Ocean islands were slaughtered and eaten until extinct (one more followed in the nineteenth century). Today, the IUCN-SSC (the World Conservation Union Species Survival Commission) classifies 38% of the world's chelonians as threatened: 4% as critically endangered, 11% as endangered and 23% as vulnerable. Chelonians, thus, contain a much higher percentage of threatened taxa than birds and mammals, of which, respectively, 11 and 25% are threatened (Baillie 1996).

Most of the international conservation efforts for chelonians are directed towards sea turtles, some towards terrestrial tortoises and a few concern freshwater species. This reflects differences in their appeal to the human psyche and differences in conservation politics rather than differences in their actual conservation needs. The high level of conservation efforts directed towards the seven magnificent sea turtle species is based on a vigorous lobby group of sea turtle researchers, managers and conservationists, the IUCN Marine Turtle Specialist Group. All seven marine turtle species are classified as threatened (IUCN 1996). This, however, does not mean that the sea turtles are, objectively, more likely to become extinct than the non-marine chelonians. It reflects the fact that the IUCN Marine Turtle Specialist Group did not classify the sea turtles according to the new, standardised IUCN criteria which were developed to assign uniform categories of threat to all the world's species and subspecies of animals and plants, whereas the IUCN Tortoise and Freshwater Turtle Specialist Group did use the standardised IUCN criteria in their status assessment of all tortoises and freshwater turtles.

This anomaly in the status assessment of sea turtles has been challenged, because there are sound, objective reasons for being able to identify species that are really on the downhill slide, from those that may be depleted but, objectively, are at present not under high risk of extinction. In discussing the "critically endangered" listing of the hawksbill turtle *Eretmochelys imbricata* (IUCN 1996), Mrosovsky (1997) pointed out that this category means "facing an extremely high risk of extinction in the wild in the immediate future" and that such an assessment should be based on verifiable data and provision of proper documentation. Webb (1997a: 10) elaborated:

"So we now have the bizarre situation where Western Australia's truly endangered swamp turtle, reduced to a handful of individuals in a single population and rightly classified as 'critically endangered' shares this prestigious position with hawksbill sea turtles, that still occupy their global distribution, are totally secure in vast areas of habitat such as in Australia, and whose global population is certainly in the hundreds of thousands, and may be in the millions. Clearly hawksbills in some countries may be having problems, but global extinction? How could it ever be achieved?"

Pritchard (1997a) speculated that even if unremitting overharvesting of sea turtles may drastically shift the "demostat" for species to a lower position, the new setting of the demostat would not be at the zero point (extinction). He suggested that even the rarest of all sea turtles, *Lepidochelys kempii*, should be identified as "conservation dependent" rather than as "endangered".

Two major international chelonian conservation conferences took place in the last few years, the results of which were published in proceedings: in 1993 at the State University of New York (Abbema 1997) and in 1995 at the Tortoise Village in Gonfaron, France (SOPTOM 1995). The 800 plus pages of chelonian papers in these two proceeding volumes combined provide an overview of status and conservation problems of the world's chelonians. The following account will mainly focus on aspects of chelonian reproduction that are important for the understanding of processes which threaten chelonian populations and which have important implication for conservation planning. If captive breeding is used in species recovery and conservation, a sound understanding of reproductive mechanisms is often needed to successfully breed an endangered species in captivity. Reproductive research can help to develop operational tools and management strategies for captive breeding. Last but not least, research into reproduction and population dynamics of wild populations can potentially impact on the health and vigour of the populations under study. These topics will be discussed in the following sections.

10.1 Environmental Degradation

Natural habitats are being lost at alarming rates around the world, affecting many chelonian species. Environmental degradation includes a variety of physical and biological changes to habitats, including structural changes, pollution and the proliferation of introduced predators and food competitors (e.g. pigs, dogs, rats and goats in the Galapagos Islands; foxes in Australia). Environmental degradation interferes in various ways with chelonian reproduction. Nesting habitats are lost or degraded by beach- and waterfront developments, or overgrown with weeds. Artificial lights close to sea turtle nesting beaches attract hatchlings away from the ocean (see Sect. 4.4). In undisturbed populations, eggs and small juveniles already show the highest predation and mortality rates (see Sect. 9.1.4). Introduced predators or native predator populations which increase due to subsidies by human activities may dramatically increase predation, in particular on eggs and juveniles. Predators include ravens which may increase dramatically in disturbed areas. Raven predation is a serious problem for chelonians on various continents, for example for desert tortoises (*Gopherus agassizii*) in California (Boarman 1997), for *Testudo kleinmanni* in Israel (Geffen and Mendelssohn 1997) and for *Pseudemydura umbrina* in Western Australia.

10.1.1 Case Study: *Pseudemydura umbrina*

A relict species, apparently little changed since the Miocene, *Pseudemydura umbrina* is the only member of its genus and has no close relatives among other members of the Chelidae (Burbidge et al. 1974). The only fossil records of *Pseudemydura* are a portion of a skull and a pygal bone from the early Miocene Riversleigh deposits of north-west Queensland, which show only slight differences from modern specimens (Gaffney et al. 1989). *Pseudemydura umbrina* has been recorded only from scattered localities in a narrow strip (3–5 km wide, 30 km long) of the Swan coastal plain with largely alluvial soils in the outskirts of Perth, Western Australia, and is one of the world's rarest and most endangered chelonians. Anecdotal information (Burbidge 1967, 1981) suggests that their stronghold was the clay soils of the Swan Valley, the first part of Western Australia developed for agriculture. Almost all this land is now cleared and either urbanised or used for intensive agriculture, airports or the extraction of clay for brick and tile manufacture.

Pseudemydura umbrina only inhabits shallow, ephemeral, winter-wet swamps on clay or sand over clay soils. These swamps are characterised by a high invertebrate species richness and biomass and a diverse flora. After the swamps fill in June or July the turtles can be found in water. *Pseudemydura umbrina* is carnivorous, eating only living food such as insect larvae, small crustaceans, worms and tadpoles. Food is only taken underwater; feeding is therefore restricted to winter and spring. When the swamps are dry in summer and autumn *P. umbrina* moves from the low swamp areas into slightly elevated bushland to aestivate in naturally occurring holes.

During the 1960s to 1980s there were two significant, monitored populations: one in Twin Swamps (155 ha) and one in Ellen Brook (65 ha) Nature Reserves, reserves which were created to protect *P. umbrina* habitat in 1962. Apart from the fact that much of the habitat outside the reserves disappeared, the remaining, reserved habitat experiences various disturbances and changes which impact on *P. umbrina* populations. Despite the protection afforded to *P. umbrina* habitat, their numbers declined from over 200 in the late 1960s to about 30 surviving wild individuals in the late 1980s (Kuchling and DeJose 1989). Immediately obvious disturbances and threats include road traffic, influx of polluted water from intensive livestock farming, increased frequency of bush fires, a lowering of the groundwater table, and introduced exotic predators like foxes, dogs and cats. Since the late 1980s, many of these problems have been addressed by management actions, e.g. fencing of habitat along roads and elsewhere to reduce road kills as well as to exclude predators; diversion of drains; fire management; and pumping of groundwater into one swamp during dry winters in order to maintain a certain water level (Burbidge and Kuchling 1994). More subtle habitat degradation includes changes of drainage patterns between swamps due to firebreak maintenance and the concomitant lowering of water levels, an increase of native avian predators and scavengers, in particular ravens (*Corvus coronoides*), due to adjacent farming operations (e.g. an intensive chicken farm), and artificial farm dams with permanent water just outside the nature reserves which attract *P. umbrina* during dry winters and springs. The turtles are then prone to predation, overheating or desiccation while trying to get through the boundary fence.

A very limited natural range, habitat loss and environmental degradation (including increased predation) seem to be the main causal factors for the decline of *P. umbrina*. The impact of environmental degradation on *P. umbrina* populations appears to occur mainly through reduced recruitment into the adult population due to an increase in the mortality of juveniles. In order to survive their first summer aestivation, it is critical for hatchlings to grow from about 5 g when they hatch in autumn to about 20 g in late spring. The length of the growing period depends on the availability of standing water in the swamps. Ravens kill juveniles and subadults, but, in contrast to foxes, do not seem to have a major impact on adult *P. umbrina*. Apart from increased mortality under environmental perturbations, egg production is under labile control mechanisms in this species (Chap. 6) and is, therefore, also easily influenced by environmental vagaries.

10.2 Exploitation

Chelonians have long been an important source of meat, eggs and traditional medicine for local people, especially in tropical and subtropical countries. In terms of biomass, turtles may be the dominant vertebrates in some freshwater habitats (Bury 1979). Single standing crop biomass estimates are among the highest reported for vertebrates, and the annual production per unit area may only be exceeded by certain fishes (Iverson 1982). Species possessing such characteristics are obvious targets for exploitation, in particular for protein-starved people in developing countries. Most countries have made little attempt to control or manage these resources effectively.

Traditional hunter–gatherer societies, for example the Australian aborigines, do not appear to deplete chelonian populations with which they have co-existed for a long time. This, however, changes when subsistence hunting is elevated due to human population growth and/or socio-economic changes. Exploitation, once it becomes unsustainable, threatens populations or species regardless of its being traditional, commercial or otherwise. Non-commercial subsistence exploitation may have significant impacts on several turtle and tortoise species or populations around the world. Today, most sea turtles and large river turtles as well as many other species are severely depleted through overexploitation.

Unfortunately, chelonians are exploited not only for human food at local subsistence levels. Sea turtles and a few other species have long been used and traded as a luxury food. Although, due to international conventions, the highly esteemed turtle soup has nearly disappeared from European and American dinner tables, a major trade in turtles and tortoises has developed over the past two decades, mainly from southern Asia into China, Hong Kong (now part of China) and Taiwan. It accelerated in the early 1990s when the billion people in China, who have always cherished turtles for food and medicine, had the financial resources to buy them in phenomenal numbers (Pritchard 1995). This food and medicine trade is seriously depleting populations in south Asia and is continuously expanding around the globe into new source countries. It has also reached the USA, from where tons of live freshwater turtles are now exported to Chinese food markets.

This shows that, even in a developed nation, scientific insight into chelonian demography and population dynamics does not necessarily lead to rational management of that resource: large *Chelydra serpentina*, the very species for which American scientists have demonstrated in decade-long research that populations cannot sustain the taking of adults and subadults (Congdon et al. 1994; Galbraith et al. 1995; Cunnington and Brooks 1996), are suddenly exported en masse to food markets in China. In some of these food markets they now, together with *Trachemys scripta*, make up the bulk of the trade (Behler 1997). The Convention on International Trade of Endangered Species (CITES) has virtually no dampening effect on the trade to the insatiable Asian food markets. This trade regularly involves endangered species listed on CITES Appendix I for which, under the convention, no commercial trade whatsoever is permissible. Many countries, even if they are signatories of CITES, do not implement effective regulatory mechanisms, which weakens the enforcement of this international law.

Trade and exploitation of chelonians is not limited to culinary or medical purposes. "Tortoise shell", for example the scutes of *Eretmochelys imbricata*, is highly sought after by the Japanese beco industry. The pet trade, although in volume much inferior to the food and medicine trade, involves many species and sometimes specifically targets endangered ones. Without any regard for the risk of extirpating the species concerned, a significant proportion of the world population of the endangered tortoise *Geochelone yniphora* was stolen in May 1996 from a breeding project in Madagascar for the illegal pet trade, and in 1997 the same happened to a breeding project of the endangered tortoise *Psammobates geometricus* in South Africa. Apart from the depletion of wild populations or the villainous demolition of years of conservation work through theft of endangered species, the lack of quarantine in the live trade of chelonians can potentially spread debilitating chelonian diseases (*Mycoplasma* sp., herpes and other viruses, bacterial and fungal pathogens and parasites) around the world and into wild populations (Jacobson 1997).

10.2.1 Case Study: *Erymnochelys madagascariensis*

Erymnochelys is a monotypic freshwater turtle genus endemic to Madagascar which is of great zoogeographic interest: it is the only living member of the family Podocnemidae to inhabit the Old World. Its closest relatives are the South American turtles of the genera *Podocnemis* and *Peltocephalus*. Subsistence hunting for local cooking pots is a major threat for the legally protected *Erymnochelys madagascariensis*. Few data on its populations are available, but the best-studied population (Kuchling 1988c) was extirpated between 1987 and 1991 by this non-commercial exploitation (Kuchling and Mittermeier 1993). In 1991 and 1992 most major *Erymnochelys* habitats and populations were assessed with a rapid survey technique which was based on interviews with local fishermen. The survey of 46 localities revealed serious depletion and local extirpation of populations due to overexploitation: 11% of the populations were considered to be "exploited but relatively good", 28% "exploited and declining", 28% "heavily exploited and depleted", 31% "possibly extinct" or "extinct", and at one locality *Erymnochelys* may have "never existed" (Kuchling 1997b). Over recent decades, the trend of *Erymnochelys* populations as a whole appears to be one of non-cyclic decline. In lakes where *Erymnochelys* was still common in the early 1970s (Tronc and Vuillemin 1973), the species was depleted or heavily depleted in the early 1990s (Kuchling and Mittermeier 1993).

Erymnochelys females reach sexual maturity with about 26 cm shell length (Kuchling 1993a) and both sexes grow to about 0.5 m shell length. Exploitation is mainly directed towards large turtles, but, due to their size, turtles are already highly appreciated as food long before reaching sexual maturity: if there are plenty of turtles only animals larger than about 18 cm shell length are taken, but any size is consumed in areas where the species is already rare. In habitats where fishing pressure suddenly increases due to an influx of people with better fishing technologies (seine nets), the extirpation pattern of *Erymnochelys* populations is characterised by an abrupt disappearance of the species: people who exploit them often have the impression that there are plenty of turtles until they are suddenly gone. The reason for this phenomenon is that, once all reproducing females are extirpated and no further recruitment of hatchlings takes place, there are still, for several years, small size classes from previous reproductive seasons growing up to a size which makes them attractive for human consumption. Once all these growing-up size classes have also been eaten, extinction is so abrupt that local people are often mystified by the sudden disappearance of the species. The smaller the size of the habitat, the faster *Erymnochelys* populations crash under exploitation pressure; under continuing exploitation, persistence at low population levels over several years is only found in very large and extended habitats (Kuchling and Mittermeier 1993).

Today, this turtle is primarily caught as a by-catch in fishing operations which have expanded dramatically over the last 15 years. Any turtle caught ends up slaughtered and consumed, although, in contrast to fish, turtles are generally not marketed because they are protected by law. The taking and killing of *E. madagascariensis* in this way seems to be the single most important threat to the survival of the species, which has already declined and disappeared from much of its former distribution. The status of *Erymnochelys* populations is progressively deteriorating. The human population of Madagascar is expected to double during the next 30 years. The concomitant need for food will further expand fishing activities. This makes the future survival of *Erymnochelys* unlikely if no conservation actions are taken.

10.3 Conservation Strategies

Approaches to conservation can be classified in different ways. One possibility is to use the biological units of concern and to distinguish between ecosystem conservation and species conservation. These two concepts are by no means mutually exclusive. In the long term, species cannot be conserved without the ecosystems they are part of. The best approach to protect species, therefore, is to protect functioning ecosystems. However, the wisdom that actions directed at single species are now oldfashioned and ineffective must be questioned, since effective conservation directed towards species as the unit of concern will also maintain ecosystems and biodiversity along the way – or it is not effective. Here I will only discuss conservation from the point of view of chelonian species or populations.

In species conservation, a distinction is commonly made between in situ conservation, meaning that a species is conserved in its natural habitat, and ex situ conservation, meaning that individuals are maintained in captivity or storage facilities (e.g. embryos or gametes frozen in liquid nitrogen). Since ex situ conservation is not concerned with the habitat of the species, its usefulness is frequently criticised by proponents of in situ and ecosystem conservation (e.g. Lever 1990; McIntyre et al. 1992). One strategic response by some members of the zoo and captive breeding community to this criticism is to blur the distinction between the terms in situ and ex situ. The term "in situ captive breeding" is now sometimes used for the situation where captive breeding (clearly an ex situ activity which, in itself, has nothing to do with preserving a species in its habitat) takes place in the species' country of origin, as opposed to facilities in countries where the species does not occur naturally (e.g. Mallinson 1991). Although, for various reasons (e.g. disease risk, transport costs, the commitment of nations to their own conservation problems), it is a recommendable approach to operate captive breeding programmes in countries where endangered species are native, I do not believe that maintaining species in cages or, to expand the matter, in bottles of liquid nitrogen in their home countries should qualify for the term "in situ". I, therefore, use the term in situ only in the restricted sense, meaning conservation of a species in its natural habitat.

Once numbers of a species become low, genetic considerations and genetic management become important for its survival chances. A large body of literature exists on this aspect (e.g. Soulé 1987), which Caughley (1994) called the small population paradigm. This approach deals with the risk of extinction as a consequence of small population size. It may involve laboratory experiments with *Drosophila*, but is mostly theoretical, uses computer modelling, applies terms such as minimal viable populations, population and habitat viability analysis, metapopulation analysis and extinction vortices, and is attractive by virtue of its seemingly "hard" scientific approach. It is beyond the scope of this book to elaborate on this topic. Another major approach to conservation, the declining population paradigm (Caughley 1994), is concerned with the external causes that drive populations towards extinction. Research is aimed at determining why populations are declining and how to neutralise the causes, but it is based on empirical field research and, sometimes, lacks scientific rigour. In this account I will largely follow this second approach, concentrating on aspects of population management that relate to reproduction, and present two case studies. Although headstarting involves captive rearing of wild progeny, I will discuss it here only in relation to its impact on wild populations.

10.3.1 Conservation of Wild Populations

The most important conservation strategy for chelonians is clearly habitat protection, which also conserves other species and natural communities. There are a few cases (e.g. sea turtle nesting beaches, Aldabra Island, Desert Tortoise Natural Area in California, nature reserves for the protection of *Pseudemydura umbrina* in Western Australia) where chelonians were the main reason to protect a habitat and where reserves are managed specifically to conserve chelonian species. In many other cases, chelonian populations profit by living in an area which happens to be protected for other reasons. A further important conservation strategy is to maintain, through management, a low level of subadult and adult mortality in chelonian populations. This can involve control of hunting, collecting, elevated predation rates and incidental mortality (e.g. road mortality, or drowning of sea turtles in trawl nets and of diamond terrapins in crab pots).

For a long time conservationists concerned about the decline of chelonian populations were intrigued by the large numbers of eggs and hatchlings of some turtles, particularly sea turtles, and the huge natural losses of these life stages through predation. An attempt to exploit this high reproductive output is headstarting, the practice of incubating eggs collected from wild nests and raising hatchlings in captivity until they reach a size that, theoretically, will protect them from the high rates of natural predation. The rationale is that these turtles will continue to enjoy high rates of survival after they are released in the wild. Headstarting is not undisputed; risks include possible nutritional deficiencies, behavioural modifications, skewed sex ratios due to unnatural incubation temperatures, diseases and lack of imprinting of natal beaches (Frazer 1997). In the USA, a headstarting programme for green (*Chelonia mydas*) and loggerhead (*Caretta caretta*) turtles operated from 1959 to 1989, but their efficiency has not yet been evaluated. Similar programmes have operated at one time or another in virtually every country were sea turtles occur (Mortimer 1995). Headstarting and release programmes also operate for several freshwater turtles, for example in India primarily for softshelled turtles (Rao 1995) and in Malaysia for *Batagur baska* and *Callagur borneoensis* (Moll 1995). Recent findings suggest that a headstarting programme for *Lepidochelys kempii* from 1978 to 1993 does show positive results and that headstarted turtles of this species can contribute to breeding populations (Pritchard 1997). The same holds true for the Hood Island Galapagos tortoise (Cayot and Morillo 1997).

Models of the population-level effects of headstarting suggest that it is unlikely to ever meet its goal of increased recruitment into the adult population without a simultaneous reduction of juvenile and adult mortality. Management efforts focusing exclusively on improving the survival in the first year of life are unlikely to be effective (Mortimer 1995; Heppel et al. 1996). Due to these models, headstarting can, however, speed up the recovery of depleted populations when adult survival is returned to and maintained at high levels.

Once wild populations have disappeared or nearly disappeared, reintroduction or population augmentations using captive bred progeny may be necessary to conserve species in their natural habitats. Dodd and Seigel (1991) reviewed relocation, repatriation and translocation (RRT) of amphibians and reptiles, including several programmes with chelonians, and concluded that the success of these conservation techniques has not yet been proven. They define success as "evidence that a self-sustaining population has been established" or that "the population is at least stable". They suggest that mere successful reproduction is insufficient. R.L. Burke (1991) points out that the persistence or extinction of RRT populations must be considered against the baseline extinction rates of similar-sized unaltered populations. In the case of the long-lived chelonians, an obvious difficulty is to establish the "long-term success" of any conservation action. The principal question remains as to whether RRT measures are a cost effective method of improving a species' chances of survival. In order for such actions to be successful, it is necessary to remove or to alleviate through management the causal factors that lead to the species' decline or disappearance from the particular area, or the measures will remain halfway technologies (Frazer 1997). At this stage, all RRT actions should be considered experimental and provisions should be made for a biologically based, long-term monitoring programme. An example of a reintroduction project can be found in the case study on *P. umbrina* (see Sect. 10.3.4).

10.3.2 Sustainable Use

Due to their importance as a resource, suggestions for managing wild chelonian populations effectively for use by local communities in tropical countries have been discussed for some time. Recent proposals include the better protection of nesting beaches in South American rivers which should lead to the recovery of river turtles of the genus *Podocnemis* and other species – and which, combined with careful management and sustained-yield cropping, should ensure their survival as a valuable resource (Mittermeier 1978; Alho 1985; Vogt 1995; Cantarelli 1997). Similar strategies were proposed for freshwater turtles in India (Moll 1991; Choudhury et al. 1997). Although the population models for North American freshwater turtles and for sea turtles (see Sect. 9.2) suggest that the harvest of subadults and adults cannot be compensated for by reducing egg and juvenile mortality alone, and particularly not in already-depleted populations, this generalisation does not necessarily apply to populations living in optimal habitats. A few chelonian populations seem to have tolerated impressive levels of exploitation, for example the olive ridley (*Lepidochelys olivacea*) colony in Oaxaca, Mexico. This colony was subjected to an annual take of many thousands of animals per year for over 20 years, yet following the final closure of the legal take, arribadas of very large and annually increasing size were observed. Furthermore, some Mediterranean tortoise populations continue to thrive despite long histories of exploitation (Pritchard 1997b). Still, more ecological and demographic data or case studies on exploited populations are needed for tropical turtles before conservation schemes that involve continuing exploitation of adult chelonians for meat should be implemented.

The concept of sustainable use of wild populations proved to be a very successful conservation scheme for crocodiles. An important usage is to harvest eggs or hatchlings from wild nests, to raise the young crocodiles in captivity and, at a certain size, to slaughter them for their meat and skin (Webb 1997b). There are certainly biological differences between crocodilians and chelonians, but, in the light of the rampant destruction of chelonian populations (as was the case with crocodile populations before the sustainable use schemes were introduced), it appears worthwhile to explore whether similar concepts could also work for chelonians. As discussed in Section 9.2, population models for various chelonians suggests that harvest (increased mortality) of adults or subadults can hardly be sustained by any chelonian population. Population stability in chelonians seems to be relatively insensitive to parameters like fecundity and nest survival, but highly sensitive to larger juvenile and adult survival (Congdon et al. 1993). This suggests that careful exploitation of eggs and hatchlings may have better prospects for sustainability than exploitation of adults. The depletion of wild populations and the insatiable demand for chelonians in east Asia suggest that ranching operations (based on collection of wild eggs) or captive breeding for commercial purposes will become increasingly viable. However, egg collection – while sparing adults – also has to be managed in a proper way. The virtual extinction of the famous leatherback sea turtle rookery at Terengannu, Malaysia, demonstrates that overexploitation of eggs for more than a generation can also cause population crashes (Limpus 1995).

No functioning sustainable use system (under which populations do not decline) of any wild chelonian population appears to be operating today: all exploitation schemes seem to be unsustainable (Galbraith et al. 1997). However, there is also a lack of data to evaluate exploitation; most conclusions are based on market surveys (not population surveys) or anecdotal information. Sustainability is not in itself a fixed entity, it is a process requiring monitoring, assessment and, if necessary, corrective actions (Webb 1997a). Despite the lack of monitoring, current levels of exploitation of chelonian populations in most locations seem to be much too great and such exploitation should be severely curtailed or banned nearly everywhere.

10.3.3 Conservation in Captivity

Once species or populations decline to a level where immediate actions are needed to avoid extinction, the transfer of individuals into captivity becomes a useful strategy. Conservation in captivity is a halfway technology which, in itself, does not address the underlying causes of population decline (Frazer 1997), but it is a means of buying time until those causes can be corrected or alleviated. Maintaining a species in captivity raises some important ethical questions. Great emphasis on conservation in captivity could cause governments and the community at large to see it as a substitute rather than a complement to conservation in the wild. A heavy reliance on human institutions for the conservation of a species is not prudent, and should always be seen as a temporary emergency strategy. All human institutions are transient and, as such, vulnerable to political vagaries. Yet another ethical concern is that species in captive situations are confined to highly protected environments. They may quickly adapt to captivity and, when released back into the milieu which caused their problems in the first place, may be even less likely to succeed. For all these reasons the period over which a species is maintained in captivity should be kept as short as possible – in particular in generation times: if the F1 generation of wild caught individuals is used for reintroduction, the problems of genetic drift can largely be avoided.

Captive breeding and release programmes have many genetic implications that should be carefully considered. In order to preserve genetic variability a large founder stock is always important. The number will obviously depend on the availability of individuals and space. For genetic reasons, random mating by pooling the breeding stock in one or a few enclosures should be avoided. The optimal breeding strategy depends on the main goal of the operation.

If the primary goal is to produce captive bred progeny for reintroduction, the breeding output of all founder individuals can be maximised and a mating strategy can be used under which mating partners rotate between breeding seasons so that, over time, each male inseminates each female. This is an attempt to involve all individuals in the production of offspring and, while maximising reproduction, to equalise founder contribution as much as possible (although sperm storage of females and sperm competition may still bias the results). In this way, a genetically diverse F1 population can be obtained, which is the sort that should be used for reintroduction.

A different breeding strategy has to be applied when a captive population has the purpose of providing a safeguard against extinction of the species in case of catastrophes impacting on wild populations; that is, when a viable captive population has to be maintained over several generations (ark paradigm). In this case it is advisable to keep the generation intervals in captivity as long as possible – which, for many chelonian species, can potentially be very long (see Sect. 9.1.4). Reproductive output should not be maximised; individuals should only produce offspring for the next breeding generation once they are getting near the end of their life (in order to slow down the generation turnover rate). Breeding pairs should remain together and mates should not be exchanged (except if one partner does not function) in order to have a reasonable pool of unrelated individuals in the F1 generation. If managed in this way, and depending on the number of founders, pairs of unrelated individuals should be available for breeding over several generations.

10.3.4 Case Study: *Pseudemydura umbrina*

The first *P. umbrina* known to science was sent to the Vienna Museum in 1839. It is not known where this specimen was collected – it was simply labelled "New Holland". The specimen remained in the Museum undescribed until 1901, when it was named by Siebenrock. No further specimens were collected until 1953 when two were found near Warbrook, 30 km north-east of the centre of the city of Perth in Western Australia. During the 1950s, field data on the species were gathered by researchers of the Western Australian Museum. In 1962 two small Class A nature reserves were created which protected some of *P. umbrina*'s remaining habitat of ephemeral clay or sand over clay swamps: Ellen Brook Nature Reserve (EBNR) of 65 ha and Twin Swamps Nature Reserve (TSNR) of 155 ha. Research into the biology of the species started in 1963 by staff and students of the University of Western Australia, and was later continued by the state wildlife authority. In the mid-1960s more than 200 *P. umbrina* were estimated to live in the two reserves (Burbidge 1967). Since then, numbers in the wild declined until the late 1980s. At TSNR numbers have dropped from over 100 in the mid-1960s (Burbidge 1967), to about 50 in the early 1970s (Burbidge 1981), to possibly three in the early 1990s. Over the same period the population at EBNR remained fairly static at around 20–30 animals. Habitat loss and environmental degradation are the main threats to the persistence of *P. umbrina* (see Sect. 10.1.1).

A captive colony of 25 *P. umbrina* was established in 1959, and its 13 surviving animals were transferred to Perth Zoo in 1964 (Kuchling et al. 1992). The first captive breeding was recorded in 1966; 26 hatchlings were produced until 1978 (Spence et al. 1979), five of which have survived until today. In 1979, the breeding stock from Perth Zoo and two females and one male of the TSNR population were transferred to the Western Australian Wildlife Research Centre of the Department of Conservation and Land Management to improve the reproductive output in captivity. Seventeen eggs were produced in 1979 and 1980, eight of which hatched. Seven hatchlings died during their first year and one in its second year. No further eggs were produced by the captive colony between 1981 and 1986. By 1987, seven males and two females of the old captive founder group survived, together with five of the offspring bred in Perth Zoo and one male, one female and one juvenile of the former TSNR population (Kuchling et al. 1992). By 1987, when I arrived in Western Australia, the wild population numbered about 30 animals in a single population. Only 17 animals remained in captivity, three of which were adult females. They had not produced eggs for 6 years. With a total world population of under 50 individuals and no success in captive breeding, *P. umbrina* was on the brink of extinction.

A 3-year crash project started in 1988 with the aim of establishing successful captive breeding techniques to increase the world number of *P. umbrina* to a more secure level (Kuchling 1988a; Kuchling and DeJose 1989). This breeding programme is still operating today and has the main aim of producing tortoises for reintroduction in order to re-establish wild populations (Kuchling et al. 1992). The captive population is kept in outdoor enclosures under similar climatic and environmental conditions as experienced by the wild populations. Breeding is optimised by carefully managing the captive environment (Kuchling and

Bradshaw 1993). This programme was and is successful (Table 10.1); today the world population of *P. umbrina* numbers again over 250 individuals, although the number of breeding adults is only about 40. The effective population size, therefore, is still critically low.

The main reason for the poor captive breeding success of *P. umbrina* pre-1987 was a lack of understanding of the reproductive biological adaptations of chelonians to unpredictable habitats (see Sect. 5.4) and of the labile control mechanisms which may operate in those species (Chap. 6). The standard freshwater turtle husbandry techniques used by zoos and breeders around the world, which are adequate to breed many other species, simply did not work for *P. umbrina*. Once these mechanisms were understood, a husbandry protocol could be established under which the species breeds with relative straightforwardness (Kuchling et al. 1992; Kuchling and Bradshaw 1993).

The Recovery Team for the Western Swamp Tortoise was founded in December 1990 to coordinate research and management of the species. Early in 1991 the last wild population of *P. umbrina* was secured against introduced predators, mainly the European Red Fox, by an electrified fence. During 1992/1993 a recovery programme for *P. umbrina* was drafted which addressed the causes for the decline of the species and which prescribed necessary conservation actions until the year 2002 (Burbidge and Kuchling 1994). Twin Swamps Nature Reserve got a fox-proof fence in early 1994 and reintroduction of captive bred F1 animals into Twin Swamps started in late 1994. By 1997, 72 captive bred hatchlings and juveniles had been released and were being monitored. A major problem at Twin Swamps is still predation by the native ravens.

In summary, the recovery actions for *P. umbrina* centre on:

- Habitat management, including predator exclusion and control, fire management, changes of drainage patterns to reconstitute former conditions, and water supplementation during dry years.
- Population monitoring and reintroduction of captive bred juveniles into managed habitat.
- Identification, protection and rehabilitation of former habitat to increase the carrying capacity of existing reserves and to establish new reserves and populations.
- Maintenance of a captive colony to produce juveniles for reintroduction and as a safeguard against problems in the wild.
- Education and awareness-raising in the Australian public about the problems of the species.

10.3.5 Case Study: *Erymnochelys madagascariensis*

In contrast to *P. umbrina*, for which habitat loss and environmental degradation are the main threats (see Sect. 10.1.1), the single major threat for *Erymnochelys madagascariensis* is exploitation at the local subsistence level (see Sect. 10.2.1). Accordingly, the approach to the conservation of this species has to be different.

Table 10.1. Results of *Pseudemydura umbrina* captive breeding project: egg and hatchling production and rearing success 1987–1997[a]

Cohorts from breeding seasons	No. of captive females laying eggs	No. of un-damaged eggs laid in captivity	No. of eggs laid and damaged in captivity[b]	No. of eggs from wild nests incubated	No. hatched	No. of hatchlings brought into captivity	No. of hatchlings or juveniles released at EBNR	No. of hatchlings released at TSNR	No. of hatchlings euthanised	No. of deaths during captive rearing	No. of juveniles >95 g released at TSNR	No. in captivity at 31.12.97
1987/88	2	7	0	0	0	0	0	0	0	0	0	0
1988/89	3	12	0	0	11	0	0	0	0	7	0	4
1989/90	5	15	1	0	11	0	0	0	0	1	10	1
1990/91	6	20	1	0	16	5	4	0	0	1	15	0
1991/92	6	19	2	0	12	1	2	0	2	1	8	4
1992/93	7	31	1	1	28	0	2	0	0	1	12	12
1993/94	8	28	2	6	30	0	0	0	1	0	12	15
1994/95	10	46	3	0	40	0	0	10	0	2	2	27
1995/96	10	48	1	0	38	0	0	0	0	2	0	36
1996/97	10	41	2	6	42	0	0	2	0	0	0	38
Total	–	267	13	13	228	6	8	12	3	15	59[c]	137

EBNR, Ellen Brook Nature Reserve; *TSNR*, Twin Swamps Nature Reserve.

[a] Eggs laid in the 1997 breeding season not included.

[b] Eggs which cracked during oviposition or recovery from nests and eggs which were improperly shelled.

[c] One juvenile was released at TSNR but returned to the captive colony with injuries and recovered. This animal is not shown under released juveniles, but under the number in captivity.

Erymnochelys is fully protected by Malagasy law and may not be hunted, killed, captured or collected without authorisation, but currently this does not prevent local exploitation and consumption of the species. None of the core habitats of the species – large rivers and lakes and their flood plains in the lowland of western Madagascar – has protected status. The species is, however, still found in some marginal habitats of one or two existing protected areas, in small rivers, creeks and lakes. The major protected area which harbours some marginal habitat is the Ankarafantsika "Strict Nature Reserve" and the adjacent Ampijoroa Forest Reserve. According to Malagasy law, hunting, fishing or any disturbance of fauna and flora is prohibited in a strict nature reserve. However, the growing human population increases pressure on the small patches of remaining intact habitats in Madagascar. Since the early 1990s local people and national migrants have moved illegally into the Ampijoroa and Ankarafantsika reserves to exploit natural resources including *Erymnochelys* whose relatively small populations declined rapidly.

In 1996, Conservation International started the 5-year Conservation and Integrated Development Project Ankarafantsika with the purpose of improving biodiversity conservation in the area. Part of these efforts is to curb the illegal exploitation of *Erymnochelys* in the reserves. With a projected reduction in adult and larger juvenile mortality through better law enforcement in the area, headstarting and the augmentation of depleted populations with captive-raised juveniles become a reasonable strategy for population recovery. The following strategic actions will commence at Ankarafantsika in 1998 (Kuchling 1997c): (1) elimination or reduction of the exploitation pressure; (2) population assessment and monitoring; (3) the production of juvenile turtles through captive breeding and rearing; and (4) reintroduction and population augmentation. These actions are designed to secure the persistence of this protected species in one protected area. If this cannot be achieved, what hope would there be to conserve *Erymnochelys* populations outside protected areas? Major additional activities will include education and sensitisation of the people in western Madagascar regarding the threats facing this turtle and a national status survey of *Erymnochelys* populations.

10.4 Commercial Captive Breeding Operations

Turtle species that have been produced commercially include soft-shelled turtles, mainly *Pelodiscus sinensis*, in ponds in Japan, China and Thailand; the red ear slider, *Trachemys scripta elegans*, and the diamondback terrapin, *Malaclemmys terrapin*, in the USA; and the green sea turtle, *Chelonia mydas*, in the Cayman Islands, Surinam, Reunion and Australia. The soft-shelled turtles and diamondback terrapin are cultured solely for food, the red ear slider for the pet industry and the green sea turtle for food, leather and decorative products (Wood 1991). At the Cayman Turtle Farm (British West Indies), F2 generations of green sea turtles have already been successfully produced in captivity (Wood and Wood 1990).

A growing number of private pet keepers are successfully breeding a number of turtle and tortoise species in captivity, and often trading their surplus through societies or the pet trade. Relatively new are commercial breeding operations of chelonians for the exotic pet trade, mainly in the USA, where Galapagos giant tortoises and spurred tortoises (*Geochelone sulcata*) are now regularly bred and traded. A similar commercial operation exists on Mauritius, where the Aldabra giant tortoise and the radiated tortoise *Geochelone radiata* are bred and marketed.

Commercial breeding operations involving chelonians are often capital-intensive and can only be viable if the products have a high enough market value, as is the case in the pet trade, the Asian food and medicine market trade, and in the production of specialities (e.g. turtle soup) for the rich man's table. Species like *Erymnochelys madagascariensis*, which are only used at the local subsistence level and which are sold locally in the same price range as chickens (Kuchling 1997b), cannot support commercial operations. Commercial breeding of chelonians is currently still limited, but may gain importance with the large-scale depletion of wild populations and the concomitant increase in their market values.

10.5 Risk Minimisation in Assessing Reproduction in Live Chelonians

Captive breeding operations – as well as the conservation management of wild populations – require the sound knowledge of reproductive data. Useful methods and techniques to assess the reproductive condition of live chelonians are presented in Chapter 2, but applying these techniques in studies of threatened and endangered species is often still a challenging, but necessary, task. Impacts of research procedures on the studied individuals and populations have to be minimal and, in order to issue the necessary permits for research, conservation officials have to be convinced that this is the case. In the following sections I will briefly discuss some risks and offer hints for reproductive research involving endangered species.

10.5.1 Risks and Limitations of Currently Used Methods

Radiography has been used for two decades to monitor reproductive output in population studies of chelonians by routinely screening females. Shelled oviducal eggs can be counted with 100% accuracy (see Sect. 2.1). The regular irradiation of females, however, is not without risks. The long-term effects of radiography on hatchling health, fecundity and survivorship of populations still have not been studied and remain unknown. This, however, does not mean that such effects cannot occur or should not be considered.

As a general rule, the life stages from gametogenesis to embryonic development are the most sensitive to irradiation. Hinton et al. (1997) hypothesised that radiography was safe for the screening of turtle populations, because development of shelled oviducal eggs is arrested in the late gastrula stage. What Hinton et al. (1997) overlooked was that not all radiographed turtles are necessarily in late gravidity, the state in which their eggs are not highly radio-sensitive. If freshwater turtle populations are studied by using terrestrial drift fences and pit falls surrounding aquatic habitats (see for example Gibbons 1990b), gravid females are regularly intercepted during their nesting movement and a reasonable proportion of females may be in the state of late gravidity. However, if females in a population are randomly captured rather than selectively captured during their late gravid state, for example in field studies of terrestrial species with dispersed nesting sites, either screening of females by radiography will not provide a complete picture of the reproductive output of a population (many females may just not be in the right condition), or individual females will need to be frequently X-rayed during the breeding season in order not to miss clutches. In such studies, there will only be a low percentage of radiographed females which are in late gravidity and carry detectable eggs which are not highly radio-sensitive. Population studies of threatened species in which individual, radio-tracked females were, or are, routinely and frequently radiographed over one or several years include studies on *Clemmys marmorata* (Scott et al. 1997), *Testudo kleinmanni* (Geffen and Mendelssohn 1991), *Psammobates geometricus* (E. Baard and M. Hofmeier, pers. comm) and *Gopherus agassizii* (Karl 1997). I consider these protocols to bear risks of adverse impacts.

Oogenesis, the formation of oocytes from oogonia, occurs throughout the reproductive life of chelonians (see Sect. 3.2.1), often after the breeding season (Altland 1951). Therefore, in contrast to mammals, including humans, in which oocyte formation is restricted to the embryonal stage, adult chelonian females are not more or less immune to X-ray damage to their germ line. Meiosis and ovulation occur just prior to the oviducal period of eggs. Fertilisation and early embryogenesis until the late gastrula stage take place in the oviducts before the eggs are fully shelled and ready for oviposition. Many females that are routinely subjected to radiography before, during or after the breeding season may be in a state in which their germ cells and embryos are highly sensitive to X-ray exposure and prone to irradiation damage. In addition, absolutely no information on their reproductive state can be gained by radiographing females which do not carry shelled eggs. Therefore, radiography is not appropriate for routine screening of reproduction in chelonian populations, and particularly not in endangered species. Lapid and Robinzon (1997) suggested a ten-fold increase in shell deformations of hatchlings of *Testudo graeca* (from 1.8 to 21.6%) after females were repeatedly X-rayed during the breeding season. Even if a causal relationship is often hard to proof scientifically, particular care should be taken if radiography is used in studies of threatened and endangered species.

Ultra-sound scanning has now been used for a decade in freshwater, terrestrial and marine turtles and is an excellent, non-invasive method to screen the reproductive conditions of females in wild populations. In contrast to radiography, which only detects shelled oviducal eggs, it provides qualitative and, in the case of small and medium sized chelonians, also quantitative data for most stages of the female reproductive cycle (see Sect. 2.2). Limitations of ultra-sound scanning include the fact that it does not allow 100% accurate counting of oviductal eggs if females carry large numbers of eggs, and that the corpora lutea, corpora albicantia or testes and epididymes cannot be readily detected.

Endoscopy has been extensively used to screen reproductive conditions in wild sea turtle populations (Limpus and Reed 1985; Owens 1997), and is also an excellent technique for studying freshwater and terrestrial chelonians. It is, however, an invasive procedure requiring anaesthesia (at least locally), sterile surgical conditions and suturing of the incisions. It provides excellent qualitative information on the reproductive condition of females, but does not generally allow the quantitative assessment of eggs or follicles. Endoscopy is the only field method which allows examination of the reproductive condition of males, since testis, epididymis and vas deferens can be visualised, which is not possible by either radiography or ultra-sound scanning. It can also be used to sex juvenile turtles. The limitations of laparoscopy include the inability to measure structures accurately; their size can only be estimated, which is difficult through wide-angle lenses.

Since endoscopy requires specialised training and equipment, and since it involves some risk to the animal, the measurement of hormone levels to determine the individual's reproductive condition is another option. In many chelonians, blood samples can be quickly and easily obtained. However, even in the well-studied sea turtles, hormone determination alone cannot provide as much information as ultra-sound scanning or endoscopy. It is, however, useful to answer some particular questions. For example, the higher testosterone levels seen in immature sea turtle males can be successfully used to sex juveniles (Owens 1997).

10.5.2 Routine Screening of Reproductive Condition

The routine screening of the reproductive condition of females in a population is an important procedure in reproductive biological and life history studies. Due to the associated risks and limitations, radiography cannot be recommended for the routine screening of reproductive condition. Endoscopy, although relatively secure when properly performed, is an invasive procedure and should not be performed frequently on the same individual. It is, however, the best available technique to investigate the gonads of juveniles (e.g. to establish sex ratios) and male chelonians. To assess and routinely screen the reproductive condition of adult females, ultra-sound scanning is clearly the first choice.

If the 100% accurate counting of oviducal eggs is imperative, females should be selected for radiography only if shelled oviductal eggs were first detected by ultrasound scanning. It is useless and potentially damaging to radiograph females in other reproductive states. Researchers generally have two film options when using radiography for field research: ready pack or cassette-type films. Cassette films have a serious drawback when working in the field: in contrast to ready pack films, dark room facilities are necessary to load, unload and store each individual sheet of cassette films. However, the drawback of ready pack films is that they need higher kvp (kilovolt peak) settings (radiation intensities) and 10–20 times longer exposure times than cassette films, which increases the radiation doses substantially (Hinton et al. 1997). In order to minimise irradiation time and intensity when X-raying endangered species, cassette films with rare earth screens should always be used in preference to ready pack films.

In species in which blood sampling can be performed without problems, hormone measurements can be straightforward, but less information is generally gained than by using endoscopy or ultra-sound scanning. An elegant, non-invasive technique is the determination of steroid hormone concentration in faecal samples of chelonians. In *Testudo hermanni* and *T. graeca*, broad correlations of faecal oestrone, epiandrosterone and corticosterone concentrations were observed with follicular growth and the testis enlargement, but no information could be gained regarding ovulation, gravidity and nesting (Gumpenberger 1996a). Thus, the information gained by faecal steroid analysis is rather limited.

Chapter 11

Conclusions and Outlook

Chelonians, at least a few species, have been used extensively in various physiological studies. Concerning reproductive physiology, a considerable number of descriptions of gonadal and endocrinological cycles are now available. At the same time, a large number of ecological and life history studies of chelonians have accumulated a substantial body of data on egg size, clutch size, clutch frequency and reproductive output for various species. However, little cross fertilisation appears to occur between the physiological and the life history approach. Until recently, few researchers explored the physiological control mechanisms of the allocation processes that determine the number of eggs and clutches, and the clutch frequency in species and populations.

The most conspicuous feature of chelonians is their box-like shell. During the assessment of reproduction in ecological and life history studies, particularly regarding freshwater and terrestrial chelonians, their box-like appearance seems to invite treating them like boxes: the external dimensions and the masses of the individual boxes are measured, and then the boxes are placed on X-ray machines to find out if there are smaller boxes (shelled eggs) inside which can be counted and measured. This black box approach – to investigate the units of reproductive output (eggs, clutches) only once they are ready to be shed, without considering the underlying patterns and control mechanisms that allow or do not allow an organism to produce them at a particular time – limits our understanding of the variability of reproduction within and between populations and years and under different environmental conditions. Although, over two decades, this method has supplied a wealth of data and insights into chelonian life histories, it has also set limits to what could be explored and understood.

Ten years ago, when I started a rescue programme for the world's most endangered chelonian species, *Pseudemydura umbrina*, the conceptual limitations of the then established physiological, ecological and life history methods became at once obvious. Despite two decades of ecological, demographic and life history research on the species (Burbidge 1967, 1981), there was still a serious lack of knowledge on aspects of its reproductive biology, knowledge which was critical for the establishment of an effective conservation programme. In order to rescue this species, which then numbered less than 50 individuals, it was imperative to gain a sound understanding of the factors controlling its reproductive output and frequency. Ultra-sound scanning proved to be an important breakthrough in advancing knowledge which is of direct relevance to chelonian reproductive management and conservation biology, and which can be used in highly endangered species without prohibitive risk.

Limitations of our current knowledge of chelonian reproduction and life history not only restrict attempts to directly improve the reproductive output and success of species and populations, for example through captive breeding. The modelling of population viability under different management scenarios (e.g. Heppel et al. 1996) is also often confounded by largely insignificant variables (e.g. number of eggs, clutches or nest survival: Congdon et al. 1993) that are quantified precisely, whereas important variables (e.g. mortality rates of different size classes of juveniles and adults) are based on guesswork. In regard to chelonian conservation biology, efforts to further increase quantitative data sets of rather insignificant variables such as egg size and clutch size, simply because they are easy to measure precisely, may be largely wasted relative to any research which helps to quantify the important variables more precisely (Webb 1997a).

Which major features characterise chelonian reproduction in addition to longevity and iteroparity? One is certainly the extensive potential for sperm storage in males and females that does not necessitate the synchronisation of male and female gonadal cycles. It enables the temporal dissociation of mating and insemination not only from spermatogenesis, but also from ovulation and fertilisation. Chelonians, thus, have a remarkable flexibility in meeting the ultimate causes for the timing of reproduction in various climates and habitats without strong selective pressure to shift or restructure basic gonadal cycles and their proximate cues. Ovulation and oviposition, energetically the most critical phases of reproduction, rely heavily on pre-programmed mechanisms (closed control) in species living in predictable environments, but labile control mechanisms often operate in species occupying unpredictable habitats. This inherent flexibility in chelonian reproduction may contribute to their – compared to lizards and snakes – conservatism regarding reproductive traits.

Some scientists warn that climatic change (global warming) may pose yet another level of threat to chelonians by impacting on sex ratios of species with temperature-dependent sex determination. I believe chelonians have experienced and survived several climatic changes over the last 200 million years, and many of them should be able to handle such situations through nest site choice or the timing of nesting, and, therefore, reproduce successfully and persist. Chelonians have already persisted through several mass extinction events in the earth's history, including the demise of the dinosaurs at the Cretaceous–Tertiary boundary. However, with the present human-caused extinction wave, chelonians are possibly at the critical juncture of their 200-million-year history. Their challenge now is to outlive their worst adversaries – humans. Many humans, on the other hand, wish to help them in their struggle. I hope that the approaches to studying and understanding chelonian reproduction presented in this book will further efforts towards their conservation.

References

Abbema J Van (1997) Proceedings: Conservation, Management and Restoration of Tortoises and Turtles, 11–16 July 1993, Purchase. New York Turtle and Tortoise Society, New York

Abrams Motz V, Callard IP (1991) Seasonal variations in oviductal morphology of the painted turtle, *Chrysemys picta*. J Morphol 207:59–71

Agassiz L (1857) Contributions to the natural history of the United States of America, vols 1, 2. Little Brown, Boston

Alho CJR (1985) Conservation and management strategies for commonly exploited Amazonian turtles. Biol Conserv 32:291–198

Alho CJR, Pádua LFM (1982) Reproductive parameters and nesting behavior of the Amazon turtle *Podocnemis expansa* (Testudinata: Pelomedusidae) in Brazil. Can J Zool 60:97–103

Allen BM (1906) The origin of the sex cells of *Chrysemys*. Anat Anz 29:217–236

Altland PD (1951) Observations on the structure of the reproductive organs of the box turtle. J Morphol 89:599–621

Altland PD, Highman B, Wood B (1951) Some effects of x-irradiation on turtles. J Exp Zool 118:1–14

Andrews RM (1982) Patterns of growth in reptiles. In: Gans C, Pough FH (eds) Biology of the Reptilia, vol 13. Academic Press, London, pp 273–320

Aschoff J (1955) Jahresperiodik der Fortpflanzung bei Warmblütern. Stud Gen 8:742–776

Auffenberg W (1964) A first record of breeding colour changes in a tortoise. J Bombay Nat Hist Soc 61:191–192

Auffenberg W, Iverson JB (1979) Demography of terrestrial turtles. In: Harless M, Morlock H (eds) Turtles perspectives and research. John Wiley, New York, pp 541–569

Auffenberg W, Khan NA (1991) Studies of Pakistan reptiles: notes on *Kachuga smithi*. Hamadryad 16:25–29

Avery HW, Vitt LJ (1984) How to get blood from a turtle. Copeia 1984:209–210

Bailey RC (1992) Why we should stop trying to measure the cost of reproduction correctly. Oikos 65:349–352

Baillie J (1996) Analysis. In: IUCN (ed) 1996 IUCN red list of threatened animals. IUCN, Gland, pp 24–41

Baker JR (1938) The evolution of breeding seasons. In: DeBeer G (ed) Evolution. Clarendon, Oxford, pp 161–177

Baker RR (1978) The evolutionary ecology of animal migration. Holmes & Meir, New York, pp 1012

Baker RR (1992) Introduction. In: Baker RR (ed) Great migrations. Weldon Owen, Sydney, pp 10–11

Bakst MR (1987) Anatomical basis of sperm-storage in the avian oviduct. Scan Microsc 1:1257–1266

Bamberg E (1975) Untersuchungen über die Spermienreifung im Nebenhoden. Habilitationsschrift Tierärztliche Hochschule, Wien

Barney R (1922) Further notes on the natural history and artificial propagation of the diamondbacked terrapin. Bull US Bur Fish 38:9–111

Barrett SL, Humphery JA (1986) Agonistic interactions between *Gopherus agassizii* (Testudinidae) and *Heloderma suspectum* (Helodermatidae). Southwest Nat 31:261–263

Beard JS (1974) Vegetation survey of Western Australia – Great Victoria Desert – explanatory notes to sheet 3. University of Western Australia Press, Nedlands

Behler JL (1997) Troubled times for turtles. In: Van Abbema J (ed) Proceedings: Conservation, Management and Restoration of Tortoises and Turtles, 11–16 July 1993, Purchase. New York Turtle and Tortoise Society, New York, pp xviii–xxii

Bellairs AA, Bryant SV (1985) Autotomy and regeneration in reptiles. In: Gans C, Billett F (eds) Biology of the Reptilia, vol 15. John Wiley, New York, pp 301–410

Bels VL, Crama YJM (1994) Quantitative analysis of the courtship and mating behavior in the loggerhead musk turtle *Sternotherus minor* (Reptilia: Kinosternidae) with comments on courtship behavior in turtles. Copeia 1994:676–684

Bennett JM (1986) A method for sampling blood from hatchling loggerhead turtles. Herpetol Rev 17:43

Berry JF, Shine R (1980) Sexual size dimorphism and sexual selection in turtles (order Testudines). Oecologia 44:185–191

Berthold P (1992) The urge to move. In: Baker RR (ed) Great migrations. Weldon Owen, Sydney, pp 14–21

Bjorndal KA, Zug GR (1995) Growth and age of sea turtles. In: Bjorndal KA (ed) Biology and conservation of sea turtles, revised edn. Smithsonian Institution Press, Washington, pp 599–600

Blüm V (1986) Vertebrate reproduction. Springer, Berlin Heidelberg New York

Boarman WI (1997) Predation on turtles and tortoises by "subsidised predators". In: Abbema J Van (ed) Proceedings: Conservation, Management and Restoration of Tortoises and Turtles, 11–16 July 1993, Purchase. New York Turtle and Tortoise Society, New York, pp 103–104

Bobyn ML, Brooks RJ (1994) Interclutch and interpopulation variation in the effects of incubation conditions on sex, survival and growth of hatchling turtles (*Chelydra serpentina*). J Zool 233:233–257

Bojanus LH (1819–1821) Anatome testudinis europaeae. Zawadzki, Vilno

Bolk L, Göppert E, Kallius E, Lubosch W (1933) Handbuch der vergleichenden Anatomie der Wirbeltiere, vol 6. Urban und Schwarzenberg, Berlin

Bolten AB, Balazs GH (1995) Biology of the early pelagic stage – the "lost year". In: Bjorndal KA (ed) Biology and conservation of sea turtles, revised edn. Smithsonian Institution Press, Washington, pp 579–581

Botte V, Angelini F (1980) Endocrine control of reproduction in reptiles: the refractory period. In: Delrio G, Brachet J (eds) Steroids and their mechanism of action in nonmammalian vertebrates. Raven Press, New York, pp 201–212

Boulon RH, Dutton PH, McDonald DL (1996) Leatherback turtles (*Dermochelys coriacea*) on St. Croix, US Virgin Islands: fifteen years of conservation. Chelon Conserv Biol 2:141–147

Bowen BW (1995) Molecular genetic studies of marine turtles. In: Bjorndal KA (ed) Biology and conservation of sea turtles, revised edn. Smithsonian Institution Press, Washington, pp 585–587

Bowen BW, Meylan AB, Avise JC (1989) An odyssey of the green sea turtle, *Chelonia mydas*: Ascension Island revisited. Proc Natl Acad Sci USA 86:573–576

Boycott RC, Bourquin O (1988) The South African tortoise book. Southern Book Publishers, Johannesburg

Bradshaw SD (1997) Homeostasis in desert reptiles. Springer, Berlin Heidelberg New York

Brannian RE (1984) A soft tissue laparotomy technique in turtles. J Am Vet Med Assoc 185:1416–1417

Brewer KJ, Ensor DM (1980a) Hormonal control of osmoregulation in the Chelonia. I. The effects of prolactin and interrenal steroids in freshwater chelonians. Gen Comp Endocrinol 42:304–309

Brewer KJ, Ensor DM (1980b) Hormonal control of osmoregulation in the Chelonia. II. The effects of prolactin and corticosterone on *Testudo graeca*. Gen Comp Endocrinol 42:304–309

Brooks RJ, Bobyn ML, Galbraith DA, Layfield JA, Nancekivell EG (1991) Maternal and environmental influence on growth and survival of embryonic and hatching snapping turtles (*Chelydra serpentina*). Can J Zool 69:2667–2676

Brown G, Brooks RJ (1994) Characteristics of and fidelity to hibernacula in a northern population of snapping turtles, *Chelydra serpentina*. Copeia 1994:222–226

Buhlmann KA (1997) Life history of chicken turtles *Deirochelys reticularia* in fluctuating aquatic habitats. In: ASIH/HL/SSAR/AFS-ELHS/AES/GIS Joint Meetings, 26 June-2 July 1997, Seattle, pp 85

Buhlmann KA, Lynch TK, Gibbons JW, Greene JL (1995) Prolonged egg retention in the turtle *Deirochelys reticularia* in South Carolina. Herpetologica 51:457–462

Bull JJ, Shine R (1979) Iteroparous animals that skip opportunities for reproduction. Am Nat 114:296–303

Bulova SJ (1997) Conspecific chemical cues influence burrow choice by desert tortoises (*Gopherus agassizii*). Copeia 1997:802–810

Burbidge AA (1967) The biology of south-western Australian tortoises. PhD Thesis University of Western Australia, Perth

Burbidge AA (1981) The ecology of the western swamp tortoise, *Pseudemydura umbrina* (Testudines, Chelidae). Aust Wildl Res 8:203–222

Burbidge AA, Kuchling G (1994) Western swamp tortoise recovery plan. Western Australian Wildlife Management Program 11, CALM, Perth

Burbidge AA, Kirsch JAW, Main AR (1974) Relationships within the Chelidae (Testudines: Pleurodira) of Australia and New Guinea. Copeia 1974:392–409

Burbidge AA, Kuchling G, Fuller PJ, Graham G, Miller D (1990) The western swamp tortoise. Western Australian Wildlife Management Program 6, Department of Conservation and Land Management, Como

Burger J (1976) Behavior of hatchling diamondback terrapins (*Malaclemys terrapin*) in the field. Copeia 1976:742–748

Burger JW (1937) Experimental sexual photoperiodicity in the male turtle *Pseudemys elegans* (Wied). Am Nat 71:481–487

Burke AC (1991) The development and evolution of the turtle body plan: inferring intrinsic aspects of the evolutionary process from experimental embryology. Am Zool 31:616–627

Burke RL (1991) Relocations, repatriations, and translocations of amphibians and reptiles: taking a broader view. Herpetologica 47:350–357

Burke RL (1993) Adaptive value of sex determination mode and hatchling sex ratio bias in reptiles. Copeia 1993:854–859

Bury RB (1979) Population ecology of freshwater turtles. In: Harless M, Morlock H (eds) Turtles perspectives and research. John Wiley, New York, pp 571–602

Bustard HR (1979) Population dynamics of sea turtles. In: Harless M, Morlock H (eds) Turtles perspectives and research. John Wiley, New York, pp 523–540

Butler BO, Graham TE (1995) Early post-emergent behavior and habitat selection in hatchling Blanding's turtles, *Emydoidea blandingii*, in Massachusetts. Chelon Conserv Biol 1:187–196

Butler BO, Bowman RD, Hull TW, Sowell S (1995) Movements and home range of hatchling and yearling gopher tortoises, *Gopherus polyphemus*. Chelon Conserv Biol 1:173–180

Butler JA, Hull TW (1996) Reproduction of the tortoise, *Gopherus polyphemus*, in northeastern Florida. J Herpetol 30:14–18

Callard GV (1975) Control of the interrenal gland of the freshwater turtle in vivo and in vitro. Gen Comp Endocrinol 25:323–331

Callard IP, Hirsch M (1976) The influence of oestradiol-17β and progesterone on the contractility of the oviduct of the turtle, *Chrysemys picta*, in vitro. J Endocrinol 68:147–152

Callard IP, Ho SM (1980) Seasonal reproductive cycles in reptiles. In: Reiter RJ, Follett BK (eds) Seasonal reproduction in higher vertebrates. Karger, Basel, pp 5–38

Callard IP, Callard GV, Lance V, Eccles S (1976) Seasonal changes in testicular structure and function and the effects of gonadotropins in the freshwater turtle, *Chrysemys picta*. Gen Comp Endocrinol 30:347–356

Callard IP, Lance V, Salhanick AR, Barad D (1978) The annual ovarian cycle of *Chrysemys picta*: correlated changes in plasma steroids and parameters of vitellogenesis. Gen Comp Endocrinol 35:245–257

Callebaut M, Van Nassauw L (1987) Demonstration by monoclonal anti-desmin of a myoid tissue coat in the preovulatory ovarian tunica albuginea of the turtle *Pseudemys scripta elegans*. Med Sci Res 15:1129–1130

Callebaut M, Van Nassauw L, Harrisson F (1997) Comparison between oogenesis and related ovarian structures in a reptile, *Pseudemys scripta elegans* (turtle) and a bird *Coturnix coturnix japonica* (quail). Reprod Nutr Dev 37:233-252

Cantarelli VH (1997) The Amazon turtles – conservation and management in Brazil. In: Van Van Abbema J (ed) Proceedings: Conservation, Management and Restoration of Tortoises and Turtles, 11-16 July 1993, Purchase. New York Turtle and Tortoise Society, New York, pp 407-410

Carpenter CC, Ferguson GW (1977) Variation and evolution of stereotyped behavior in reptiles. In: Gans C, Tinkle DW (eds) Biology of the Reptilia, vol 7. Academic Press, London, pp 335-554

Carr A (1952) Handbook of turtles. Cornell University Press, Ithaca

Carr A (1967) So excellent a fish: a natural history of sea turtles. Scribner, New York

Carr A (1987) New perspectives on the pelagic stage of sea turtle development. Conserv Biol 1:103-121

Carr A (1995) Notes on the behavioral ecology of sea turtles. In: Bjorndal KA (ed) Biology and conservation of sea turtles, revised edn. Smithsonian Institution Press, Washington, pp 19-26

Carr A, Coleman PJ (1974) Seafloor spreading theory and the odyssey of the green turtle. Nature 249:128-130

Carr A, Hirth H (1961) Social facilitation in green turtle siblings. Anim Behav 9:68-70

Casares M (1995) Untersuchungen zum Fortpflanzungsgeschehen bei Riesenschildkröten (*Geochelone elephantopus* und *G. gigantea*) und Landschildkröten (*Testudo graeca* und *T. hermanni*) anhand von Ultraschalldiagnostik und Steroidanalyse im Kot. Zool Gart NF 65:50-76

Caspers GJ, Reinders GJ, Leunissen JAM, Wattel J, Dejong WW (1996) Protein sequences indicate that turtles branched off from the amniote tree after mammals. J Mol Evol 42:580-586

Caughley G (1994) Directions in conservation biology. J Anim Ecol 63:215-244

Cayot L, Morillo G (1997) Rearing and repatriation of Galapagos tortoises: *Geochelone nigra hoodensis*, a case study. In: Van Abbema J (ed) Proceedings: Conservation, Management and Restoration of Tortoises and Turtles, 11-16 July 1993, Purchase. New York Turtle and Tortoise Society, New York, pp 178-183

Chaikoff IL, Entenman C (1946) The lipids of blood, liver and egg yolk in the turtle. J Biol Chem 166:683-689

Chaloupka MY, Musick JA (1997) Age, growth, and population dynamics. In: Lutz PL, Musick JA (eds) The biology of sea turtles. CRC Press, Boca Raton, pp 233-276

Charnov EL, Bull JJ (1977) When is sex environmentally determined? Nature 266:828-830

Chessman BC (1978) Ecological studies of freshwater turtles in south-eastern Australia. PhD Thesis Monash University, Melbourne

Choudhury BC, Bhupathy S, Moll EO (1997) Conservation and management of freshwater turtles and land tortoises in India. In: Van Abbema J (ed) Proceedings: Conservation, Management and Restoration of Tortoises and Turtles, 11-16 July 1993, Purchase. New York Turtle and Tortoise Society, New York, pp 301

Choudhury S, De TK, Maiti BR, Gosh A (1982) Circadian rhythm in blood sugar and adrenomedullary hormonal concentrations in an avian and a reptilian species. Gen Comp Endocrinol 46:110-112

Christiansen JL, Burken RR (1979) Growth and maturity of the snapping turtle (*Chelydra serpentina*) in Iowa. Herpetologica 35:261-266

Christiansen JL, Dunham AE (1972) Reproduction of the yellow mud turtle (*Kinosternon flavescens flavescens*) in New Mexico. Herpetologica 28:130-137

Churchill TA, Storey KB (1992) Natural freezing survival by painted turtles *Chrysemys picta marginata* and *C. picta bellii*. Am J Physiol 262:R530-R537

Cichon M (1997) Evolution of longevity through optimal resource allocation. Proc R Soc Lond B 264:1383-1388

Clay BT (1981) Observations on the breeding biology and behaviour of the long-necked tortoise *Chelodina onlonga*. J R Soc West Aust 4:27-32

Cloudsley-Thompson JL (1970) On the biology of the desert tortoise *Testudo sulcata* in Sudan. J Zool 160:17–33

Cloudsley-Thompson JL (1982) Rhythmic activity in young red-eared terrapins (*Pseudemys scripta elegans*). Br J Herpetol 6:188–194

Cogger HG (1988) Reptiles and amphibians of Australia, 4th edn. Reed, Sydney

Combescot C (1955a) Sexualité et cycle génital de la tortue d'eau algérienne, *Emys leprosa* Schw. Bull Soc Hist Nat Afr N 45:366–377

Combescot C (1955b) Action de l'hypophysectomie sur les éléments testiculaires de la tortue d'eau algérienne, *Emys leprosa* Schw. Bull Soc Hist Nat Afr N 46:30–32

Combescot C (1955c) Action de l'hypophysectomie sur les éléments testiculaires d'une tortue terrestre, *Testudo ibera* Pal. Bull Soc Hist Nat Afr N 46:98–99

Congdon JD, Gatten RE (1989) Movements and energetics of nesting *Chrysemys picta*. Herpetologica 45:94–100

Congdon JD, Gibbons JW (1990a) The evolution of turtle life histories. In: Gibbons JW (ed) The life history and ecology of the slider turtle. Smithsonian Institution Press, Washington, DC, pp 45–54

Congdon JD, Gibbons JW (1990b) Turtle eggs: their ecology and evolution. In: Gibbons JW (ed) The life history and ecology of the slider turtle. Smithsonian Institution Press, Washington, DC, pp 109–123

Congdon JD, Tinkle DW (1982) Reproductive energetics of the painted turtle (*Chrysemys picta*). Herpetologica 38:228–237

Congdon JD, Tinkle DW, Breitenbach GL, Loben Sels RC van (1983) Nesting ecology and hatching success in the turtle *Emydoidea blandingii*. Herpetologica 39:417–429

Congdon JD, Breitenbach GL, Loben Sels RC van, Tinkle DW (1987) Reproduction and nesting ecology of snapping turtles (*Chelydra serpentina*) in southeastern Michigan. Herpetologica 43:39–54

Congdon JD, Dunham AE, Loben Sels RC van (1993) Delayed sexual maturity and demographics of Blanding's turtle (*Emydoidea blandingii*): implications for conservation and management of long-lived organisms. Conserv Biol 7:826–833

Congdon JD, Dunham AE, Loben Sels RC van (1994) Demographics of common snapping turtles (*Chelydra serpentina*): implications for conservation and management of long-lived organsims. Am Zool 34:397–408

Cooper WE Jr, Greenberg N (1992) Reptilian coloration and behavior. In: Gans C, Crews D (eds) Biology of Reptilia, vol 18. Univ Chicago Press, Chicago, pp 298–422

Costanzo JP, Iverson JB, Wright MF, Lee RE (1995) Cold hardiness and overwintering strategies of hatchlings in an assemblage of northern turtles. Ecology 76:1772–1785

Courty Y, Morel F, Dufaure JP (1987) Characterization and androgenic regulation of major mRNAs coding for epididymal proteins in a lizard (*Lacerta vivipara*). J Reprod Fertie 81:443–451

Crastz F (1982) Embryological stages of the marine turtle *Lepidochelys olivacea* (Eschscholtz). Rev Biol Trop 30:113–120

Cree A, Cockrem JF, Brown MA, Watson PR, Guillette LJ Jr, Newman DG, Chambers GK (1991) Laparoscopy, radiography, and blood analyses as techniques for identifying the reproducive condition of female Tuatara. Herpetologica 47:238–249

Cree A, Cockrem JF, Guillette LJ Jr (1992) Reproductive cycles of male and female Tuatara (*Sphenodon punctatus*) on Stephens Island, New Zealand. J Zool Lond 226:199–217

Crews D (1989) Unisexual organisms as model systems for research in the behavioral neurosciences. In: Dawley RM, Bogart JP (eds) Evolution and ecology of unisexual vertebrates. New York State Museum, Albany, pp 132–143

Crews D (1996) Temperature-dependent sex determination – the interplay of steroid hormones and temperature. Zool Sci 13:1–13

Crews D, Gans C (1992) The interaction of hormones, brain, and behavior: an emerging discipline in herpetology. In: Gans C, Crews D (eds) Biology of Reptilia, vol 18. Univ Chicago Press, Chicago, pp 1–23

Crews D, Licht P (1975) Stimulation of in vitro steroid production in turtle ovarian tissue by reptilian, amphibian, and mammalian gonadotropins. Gen Comp Endocrinol 27:71–83

Crews D, Cantu AR, Bergeron JM (1996) Temperature and non-aromatizable androgens – a common pathway in male sex determination in a turtle with temperature-dependent sex determination. J Endocrinol 149:457–463

Cunnington DC, Brooks RJ (1996) Bet-hedging theory and eigenelasticity: a comparison of the life histories of loggerhead sea turtles (*Caretta caretta*) and snapping turtles (*Chelydra serpentina*). Can J Zool 74:291–296

Cyrus RV, Mahmoud IY, Klicka J (1978) Fine structure of the corpus luteum of the snapping turtle *Chelydra serpentina*. Copeia 1978:622–627

Darwin C (1871) The descent of man, and selection in relation to sex (2 vols). Appleton, New York

De TK, Maiti BR (1989) Study of the genital tract during the annual testicular cycle of the soft-shelled turtle, *Lissemys punctata punctata* (Lacépède). Zool Anz 223:116–123

Debraga M, Rieppel O (1997) Reptile phylogeny and the interrelationships of turtles. Zool J Linn Soc 120:281–354

Dingle H (1992) Types of migration. In: Baker RR (ed) Great migrations. Weldon Owen, Sydney, pp 22–29

Doak DF, Kareiva P, Klepetka B (1994) Modeling population viability for the desert tortoise in the western Mojave Desert. Ecol Appl 4:446–460

Dodd CK Jr (1997) Population structure and the evolution of sexual size dimorphism and sex ratios in an insular population of Florida box turtles (*Terrapene carolina bauri*). Can J Zool 75:1495–1507

Dodd CK Jr, Seigel RA (1991) Relocation, repatriation, and translocation of amphibians and reptiles: are they conservation strategies that work? Herpetologica 47:336–350

Dorizzi M, Richardmercier N, Pieau C (1996) The ovary retains male potential after the thermosensitive period for sex determination in the turtle *Emys orbicularis*. Differentiation 60:193–201

Drent RH, Daan S (1980) The prudent parent: energetic adjustments in avian breeding. Ardea 68:225–252

Dubois W, Pudney J, Callard IP (1988) The annual testicular cycle in the turtle, *Chrysemys picta*: a histochemical and electron microscopic study. Gen Comp Endocrinol 71:191–204

Duda PL, Gupta VK (1982) Transabdominal migration of ova in some freshwater turtles. Proc Indian Acad Sci 91:189–197

Duvall D, Guillette LJ Jr, Jones RE (1982) Environmental control of reptilian reproductive cycles. In: Gans C, Pough FH (eds) Biology of the Reptilia, vol 13. Academic Press, London, pp 201–231

Ehrenfeld DW (1979) Behavior associated with nesting. In: Harless M, Morlock H (eds) Turtles perspectives and research. John Wiley, New York, pp 417–434

Elgar MA, Heaphy LJ (1989) Covariation between clutch size, egg weight and shape: comparative evidence for chelonians. J Zool 219:137–152

Engen S, Sæther BE (1994) Optimal allocation of resources to growth and reproduction. Theor Popul Biol 46:232–248

Ernst CH, Barbour RW (1989) Turtles of the world. Smithsonian Institution Press, Washington

Ernst CH, Zug GR (1994) Observations on the reproductive biology of the spotted turtle, *Clemmys guttata*, in southeastern Pennsylvania. J Herpetol 28:99–102

Ernst CH, Barbour RW, Lovich JE (1994) Turtles of the United States and Canada. Smithsonian Institution Press, Washington

Etchberger CR, Ewert MA, Nelson CE (1990) The effects of incubation temperature on hatching success, survival and growth in *Graptemys kohni*. Am Zool 30:56A

Etches RJ, Petitte JN (1990) Reptilian and avian follicular hierarchies: models for the study of ovarian development. J Exp Zool Suppl 4:112–122

Evans SE (1988) The early history and relationships of the Diapsida. In: Benton MJ (ed) The phylogeny and classification of tetrapods, vol 1: amphibians, reptiles, birds. Clarendon Press, Oxford, pp 221–260

Ewert MA (1979) The embryo and its egg: development and natural history. In: Harless M, Morlock H (eds) Turtles perspectives and research. John Wiley, New York, pp 333–413

Ewert MA (1985) Embryology of turtles. In: Gans C, Billet F, Maderson P (eds) Biology of the Reptilia, vol 14. John Wiley, New York, pp 77–267

Ewert MA (1991) Cold torpor, diapause, delayed hatching and aestivation in reptiles and birds. In: Deeming DC, Ferguson MWJ (eds) Egg incubation: its effects on embryonic development in birds and reptiles. Cambridge Univ Press, Cambridge, pp 173–191

Ewert MA, Nelson CE (1991) Sex determination in turtles: divers patterns and some possible adaptive values. Copeia 1991:50–69

Ewert MA, Jackson DR, Nelson CE (1994) Patterns of temperature-dependent sex determination in turtles. J Exp Zool 270:3–15

Ewing HE (1933) Reproduction in the eastern box turtle *Terrapene carolina carolina* (Linne). Copeia 1933:95–96

Figler RA, MacKenzie DS, Owens DW, Licht P, Amoss MS (1989) Increased levels of arginin vasotocin and neurophysin during nesting in sea turtles. Gen Comp Endocrinol 73:223–232

Fiorindo RP (1980) Further evidence for a prolactin-stimulating neurohormone in reptiles. Gen Comp Endocrinol 40:52–58

Fischer K (1964) Sonnenkompaßorientierung und spontane Rischtungstendenz bei jungen Suppenschildkröten (*Chelonia mydas* L.). Verh Dtsch Zool Ges 1964:546–556

Fischer K (1974) Die Steuerung der Fortpflanzungszyklen bei männlichen Reptilien. Fortschr Zool 22:362–390

Fitch HS (1981) Sexual size differences in reptiles. Misc Publ Mus Nat Hist Univ Kansas 70:1–72

Flores-Villela OA, Zug GR (1995) Reproductive biology of the chopontil, *Claudius angustatus* (Testudines: Kinosternidae), in southern Veracruz, Mexico. Chelon Conserv Biol 1:181–186

Follett BK (1984) Birds. In: Lamming GE (ed) Marshall's physiology of reproduction, vol 1: reproductive cycles of vertebrates. Churchill Livingstone, Edinburgh, pp 283–350

Foote RW (1978) Nesting of *Podocnemis unifilis* (Testudines: Pelomedusidae) in the Colombian Amazon. Herpetologica 34:333–339

Forsman A, Shine R (1995) Sexual size dimorphism in relation to frequency of reproduction in turtles (Testudines: Emydidae). Copeia 1995:727–729

Fox H (1977) The urogenital system of reptiles. In: Gans C, Parsons TS (eds) Biology of the Reptilia, vol 6. Academic Press, London, pp 1–157

Frazer NB (1997) Turtle conservation and halfway technology: what is the problem? In: Abbema J Van (ed) Proceedings: Conservation, Management and Restoration of Tortoises and Turtles, 11–16 July 1993, Purchase. New York Turtle and Tortoise Society, New York, pp 422–225

Frazer NB, Gibbons JW, Green JL (1990) Life tables of a slider turtle population. In: Gibbons JW (ed) The life history and ecology of the slider turtle. Smithsonian Institution Press, Washington, DC, pp 183–200

Frazer NB, Gibbons JW, Green JL (1991) Life history and demography of the common mud turtle *Kinosternon subrubrum* in South Carolina, USA. Ecology 72:2218–2231

Fritz U (1991) Balzverhalten und Systematik in der Subtribus Nectemydina 2. Vergleich oberhalb des Artniveaus und Anmerkungen zur Evolution. Salamandra 27:129–142

Fritz U, Obst FJ (1996) Zur Kenntnis der Celebes-Erdschildkröte, *Heosemys yuwonoi* (McCord, Iverson & Boeadi, 1995). Herpetofauna 18:27–34

Furieri P (1959) La secretione dell'epididimo e del rene sessuale nei rettili studio comparativo. Boll Zool 26:457–474

Gadow HJ (1886) The reproduction of the carapax in tortoises. J Anat Physiol 20:220–224

Gaffney ES (1984) Historical analysis of theories of chelonian relationship. Syst Zool 33:383–401

Gaffney ES, Meeker LJ (1983) Skull morphology of the oldest turtles: preliminary description of *Proganochelys quenstedti*. J Vertebr Paleontol 3:25–28

Gaffney ES, Meylan PA (1988) A phylogeny of turtles. In: Benton MJ (ed) The phylogeny and classification of tetrapods, vol 1: amphibians, reptiles, birds. Clarendon Press, Oxford, pp 157–219

Gaffney ES, Archer M, White A (1989) Chelid turtles from the Miocene freshwater limestones of Riversleigh Station, northwestern Queensland, Australia. Am Mus Novit 2959:1–10

Gaffney ES, Meylan PA, Wyss AR (1991) A computer assisted analysis of the relationships of the higher categories of turtles. Cladistics 7:313–335

Galbraith DA, Brooks RJ, Obbard ME (1989) The influence of growth rate on age and body size at maturity in female snapping turtles (*Chelydra serpentina*). Copeia 1989:896–904

Galbraith DA, Brooks RJ, Brown GP (1997) Can management intervention achieve sustainable exploitation of turtles? In: Abbema J Van (ed) Proceedings: Conservation, Management and Restoration of Tortoises and Turtles, 11–16 July 1993, Purchase. New York Turtle and Tortoise Society, New York, pp 186–194

Ganzhorn D, Licht P (1983) Regulation of seasonal gonadal cycles by temperature in the painted turtle, *Chrysemys picta*. Copeia 1983:347–358

Gapp DA, Ho SM, Callard IP (1979) Plasma levels of vitellogenin in *Chrysemys picta* during the annual gonadal cycle: measurement by specific radioimmunoassay. Endocrinology 104:784–790

Garsith A, Sidis I (1985) Sexual activity in the terrapin, *Mauremys caspica rivulata*, in Israel in relation to the testicular cycle and climatic factors. J Herpetol 19:254–260

Garstka WR, Gross M (1990) Activation and inhibition of sperm motility by kidney products in the turtle, *Trachemys scripta*. Comp Biochem Physiol 95A:329–335

Gauthier JA, Kluge AG, Rowe T (1988) The early evolution of the Amniota. In: Benton MJ (ed) The phylogeny and classification of tetrapods, vol 1: amphibians, reptiles, birds. Clarendon Press, Oxford, pp 103–155

Geffen E, Mendelssohn H (1991) Preliminary study on the breeding pattern of the Egyptian tortoise, *Testudo kleinmanni*, in Israel. Herpetol J 1:574–577

Geffen E, Mendelssohn H (1997) Avian predation on tortoises in Israel. In: Abbema J Van (ed) Proceedings: Conservation, Management and Restoration of Tortoises and Turtles, 11–16 July 1993, Purchase. New York Turtle and Tortoise Society, New York, pp 105

Georges A (1983) Reproduction of the Australian freshwater turtle *Emydura krefftii* (Chelonia: Chelidae). J Zool 201:331–350

Georges A (1984) Observations on the nesting and natural incubation of the long-necked tortoise *Chelodina expansa* in south-east Queensland. Herpetofauna 15:27–31

Georges A (1985) Reproduction and reduced body size of reptiles in unproductive insular environments. In: Grigg G, Shine R, Ehmann H (eds) Biology of Australasian frogs and reptiles. Royal Zool Soc New South Wales, Mosman, pp 311–318

Georges A (1988) Sex determination is independent of incubation temperature in another chelid turtle, *Chelodina longicollis*. Copeia 1988:248–254

Georges A (1992) Thermal characteristics and sex determination in field nests of the pig-nosed turtle, *Carettochelys insculpta* (Chelonia: Carettochelydidae), from northern Australia. Aust J Zool 40:511–521

Georges A, Limpus CJ, Stoutjesdijk R (1994) Hatchling sex in the marine turtle *Caretta caretta* is determined by proportion of development at a temperature, not daily duration of exposure. J Exp Zool 270:432–444

Giannoukos G, Callard IP (1996) Radioligand and immunochemical studies of turtle oviduct progesterone and estrogen receptors: correlations with hormone treatment and oviduct contractility. Gen Comp Endocrinol 101:63–75

Gibbons JW (1990a) The life history and ecology of the slider turtle. Smithsonian Institution Press, Washington, DC

Gibbons JW (1990b) Turtle studies at SREL: a research perspective. In: Gibbons JW (ed) The life history and ecology of the slider turtle. Smithsonian Institution Press, Washington, DC, pp 19–44

Gibbons JW, Green JL (1978) Selected aspects of the ecology of the chicken turtle, *Deirochelys reticularia* (Latreille) (Reptilia, Testudines, Emydidae). J Herpetol 12:237–241

Gibbons JW, Green JL (1979) X-ray photography: a technique to determine reproductive patterns of freshwater turtles. Herpetologica 35:86–89

Gibbons JW, Green JL (1990) Reproduction in the slider and other species of turtles. In: Gibbons JW (ed) The life history and ecology of the slider turtle. Smithsonian Institution Press, Washington, DC, pp 124–134

Gibbons JW, Lovich JE (1990) Sexual dimorphism in turtles with emphasis on the slider turtle (*Trachemys scripta*). Herpetol Monogr 4:1–29

Gibbons JW, Nelson DH (1978) The evolutionary significance of delayed emergence from the nest by hatchling turtles. Evolution 32:297–303

Gibson CWD, Hamilton J (1983) Feeding ecology and seasonal movements of giant tortoises on Aldabra Atoll. Oecologia 56:84–92

Girondot M, Fretey J (1996) Leatherback turtles, *Dermochelys coriacea*, nesting in French Guiana, 1978–1995. Chelon Conserv Biol 2:204–208

Gist DH, Jones JM (1989) Sperm storage within the oviduct of turtles. J Morphol 199:379–384

Gist DH, Hess RA, Thurston RJ (1992) Cytoplasmic droplets of painted turtle spermatozoa. J Morphol 214:153–158

Goode J (1965) Nesting behavior of freshwater tortoises in Victoria. Victorian Nat 82:218–222

Goode J (1967) Freshwater tortoises of Australia and New Guinea (in the family Chelidae). Lansdowne Press, Melbourne

Goss R (1969) Principles of regeneration. Academic Press, New York

Gould SJ, Vrba ES (1982) Exaptation – a missing term in the science of form. Paleobiology 8:4–15

Gourley E (1972) Circadian activity rhythm of the gopher tortoise (*Gopherus polyphemus*). Anim Behav 20:13–20

Gourley E (1979) Rhythms. In: Harless M, Morlock H (eds) Turtles perspectives and research. John Wiley, New York, pp 509–520

Graham TE (1979) Life history techniques. In: Harless M, Morlock H (eds) Turtles perspectives and research. John Wiley, New York, pp 73–95

Graham TE, Petikas PJ (1989) Correcting for magnification when taking measurements directly from radiographs. Herpetol Rev 20:46–47

Gregory LF, Gross TS, Bolten AB, Bjorndal KA, Guillette LJ Jr (1996) Plasma corticosterone concentrations associated with acute captivity stress in wild loggerhead sea turtles (*Caretta caretta*). Gen Comp Endocrinol 104:312–320

Gross TS, Crain DA, Bjorndal KA, Bolten AB, Carthy RR (1995) Identification of sex in hatchling loggerhead turtles (*Caretta caretta*) by analysis of steroid concentrations in chorioallantoic/ amniotic fluid. Gen Comp Endocrinol 99:204–210

Guillette LJ Jr, Bjorndal KA, Bolten A, Gross T, Palmer B, Witherington B, Matter J (1991) Plasma estradiol-17β, progesterone, prostaglandin F, and prostaglandin E_2 concentrations during natural oviposition in the loggerhead turtle (*Caretta caretta*). Gen Comp Endocrinol 82:121–130

Gumpenberger C (1996a) Steroidhormongehalt in Kot von Griechischen und Maurischen Landschildkröten im Jahresgang und Kontrolle der Ovarien mit Ultraschalluntersuchungen. Doctor med vet Thesis Veterinärmedizinische Universität Wien, Wien

Gumpenberger M (1996b) Untersuchungen am Harntrakt und weiblichen Genitaltrakt von Schildkröten mit Hilfe bildgebender Diagnostik. Doctor med vet Thesis Veterinär-medizinische Universität Wien, Wien

Gupta VK (1987) Retention of eggs by the emydine turtles *Kachuga tectum tectum* and *Katchuga smithi*. J Bombay Nat Hist Soc 84:445–447

Guraya SS (1989) Ovarian follicles in reptiles and birds. Springer, Berlin Heidelberg New York

Guyot G, Pieau C, Renous S (1994) Developement embryonnaire d'une tortue terrestre, la tortue d'hermann, *Testudo hermanni* Gmelin, 1789. Ann Sci Nat Zool 15:115–137

Gwinner E (1986) Circannual rhythms. Springer, Berlin Heidelberg New York

Gyuris E (1993) Factors that control the emergence of green turtle hatchlings from the nest. Wildl Res 20:345–353

Hailey A, Loumbourdis NS (1988) Egg size and shape, clutch dynamics, and reproductive effort in European tortoises. Can J Zool 66:1527–1536

Hailey A, Loumbourdis NS (1990) Population ecology and conservation of tortoises: demo-graphic aspects of reproduction in *Testudo hermanni*. Herpetol J 1:425–434

Hailmann JP, Elowson AM (1992) Ethogram of the nesting female loggerhead (*Caretta caretta*). Herpetologica 48:1–30

Hall TR, Chadwick A, Scanes CG, Callard IP (1978) Effects of hypothalamic extract and steroids on secretion of prolactin, growth hormone and luteinizing hormone from pituitaries of *Chrysemys picta* cultured in vitro. J Endocrinol 76:169–170

Harless M (1979) Social behavior. In: Harless M, Morlock H (eds) Turtles perspectives and research. John Wiley, New York, pp 475–492

Harless M, Morlock H (1979) Turtles perspectives and research. John Wiley, New York

Hattan LR, Gist DH (1975) Seminal receptacles in the eastern box turtle, *Terrapene carolina*. Copeia 1975:505–510

Heck J, Mackenzie DS, Rostal D, Medler K, Owens D (1997) Estrogen induction of plasma vitellogenin in the Kemp's ridley sea turtle (*Lepidochelys kempi*). Gen Comp Endocrinol 107:280–288

Hendrickson JR (1958) The green turtle, *Chelonia mydas* (Linn.) in Malaya and Sarawak. Proc Zool Soc Lond 130:455–535

Henen BT (1997) Seasonal and annual energy budgets of female desert tortoises (*Gopherus agassizii*). Ecology 78:283–296

Heppell SS, Crowder LB, Crouse DT (1996) Models to evaluate headstarting as a management tool for long-lived turtles. Ecol Appl 6:556–565

Highfield AC (1996) Practical encyclopedia of keeping and breeding tortoises and freshwater turtles. Carapace Press, London

Hinton TG, Fledderman PD, Lovich JE, Congdon JD, Gibbons JW (1997) Radiographic determination of fecundity: is the technique safe for developing turtle embryos? Chelon Conserv Biol 2:409–414

Ho SM (1987) Endocrinology of vitellogenesis. In: Norris DO, Jones RE (eds) Hormones and reproduction in fishes, amphibians, and reptiles. Plenum Press, New York, pp 145–169

Ho SM, Litalian J, Callard IP (1980) Studies on reptilian yolk, *Chrysemys* vitellogenin and phosvitin. Comp Biochem Physiol 65B:139–144

Ho SM, Wangh LJ, Callard IP (1985) Sexual differences in the in vitro induction of vitellogenesis in the turtle: role of pituitary and growth hormone. Comp Biochem Physiol 81B:467–472

Hubert J (1985) Origin and development of oocytes. In: Gans C, Billet F, Maderson P (eds) Biology of the Reptilia, vol 14. John Wiley, New York, pp 41–74

Hutton KE (1960) Seasonal physiological changes in the red-eared turtle, *Pseudemys scripta elegans*. Copeia 1960:360–362

IUCN (1996) 1996 IUCN red list of threatened animals. IUCN, Gland

Iverson JB (1982) Biomass in turtle populations: a neglected subject. Oecologia 55:69–76

Iverson JB (1990) Nesting and parental care in the mud turtle, *Kinosternon flavescens*. Can J Zool 68:230–233

Iverson JB (1991) Life history and demography of the yellow mud turtle, *Kinostrenon flavescens*. Herpetologica 47:373–395

Iverson JB (1992a) A revised checklist with distribution maps of the turtles of the world. Privately printed, Richmond

Iverson JB (1992b) Correlates of reproductive output in turtles (order Testudines). Herpetol Monogr 6:25–42

Iverson JB, Balgooyen CP, Byrd KK, Lyddan KK (1993) Latidudinal variation in egg and clutch size in turtles. Can J Zool 71:2448–2461

Jacobson ER (1997) Diseases in wild populations of turtles and tortoises: the chelonian charisma vs. coincidence conundrum. In: Abbema J Van (ed) Proceedings: Conservation, Management and Restoration of Tortoises and Turtles, 11–16 July 1993, Purchase. New York Turtle and Tortoise Society, New York, pp 87–90

Jacobson ER, Schumacher J, Green M (1992) Field and clinical techniques for sampling and handling blood for hematologic and selected biochemical determinations in the desert tortoise, *Xerobates agassizii*. Copeia 1992:237–241

Janzen FJ (1993a) The influence of incubation temperature and family on eggs, embryos, and hatchlings of the smooth softshell turtle (*Apalone mutica*). Physiol Zool 66:349–373

Janzen FJ (1993b) An experimental analysis of natural selection on body size of hatchling turtles. Ecology 74:332–341

Janzen FJ, Paukstis GL (1991) Environmental sex determination in reptiles: ecology, evolution and experimental design. Q Rev Biol 66:149–179

Jarling C, Scaperi M, Bleichert A (1989) Circadian rhythm in the temperature preference of the turtle, *Chrysemys* (= *Pseudemys*) *scripta elegans*, in a thermal gradient. J Therm Biol 14:173–178

Jeyasuria P, Roosenburg WM, Place AR (1994) Role of P-450 aromatase in sex determination of the diamondback terrapin, *Malaclemmys terrapin*. J Exp Zool 270:95–111

Jones RE (1987) Ovulation: insights about the mechanism based on a comparative approach. In: Norris DO, Jones RE (eds) Hormones and reproduction in fishes, amphibians, and reptiles. Plenum Press, New York, pp 203–240

Junk WJ, Silva VMF da (1997) Mammals, reptiles and amphibians. In: Junk WJ (ed) The central Amazon floodplain. Springer, Berlin Heidelberg New York, pp 409–417

Karl AE (1997) Reproductive strategies of the desert tortoise *Gopherus agassizii* in the eastern Mojave Desert. In: ASIH/HL/SSAR/AFS-ELHS/AES/GIS Joint Meetings, 26 June–2 July 1997, Seattle, 180 pp

Keller C, Diaz-Pangiagua C, Andreu AC (1997) Post-emergent field activity and growth rates of hatchling spur-thighed tortoises, *Testudo graeca*. Can J Zool 75:1089–1098

Kennett R (1996) Growth models for two species of freshwater turtle, *Chelodina rugosa* and *Elseya dentata*, from the wet-dry tropics of northern Australia. Herpetologica 52:383–395

Kennett RM, Georges A (1990) Habitat utilization and its relationship to growth and reproduction of the eastern long-necked turtle, *Chelodina longicollis* (Testudinata: Chelidae), from Australia. Herpetologica 46:22–33

Kennett RM, Christian K, Pritchard D (1993a) Underwater nesting by the tropical freshwater turtle, *Chelodina rugosa* (Testudinata: Chelidae). Aust J Zool 41:47–52

Kennett RM, Georges A, Palmer-Allen M (1993b) Early developmental arrest during immersion of eggs of a tropical freshwater turtle, *Chelodina rugosa* (Testudinata: Chelidae), from northern Australia. Aust J Zool 41:37–45

Kim SH, Cho KW, Koh GY (1987) Circannual changes in renin concentration, plasma electrolytes, and osmolality in the freshwater turtle. Gen Comp Endocrinol 67:383–389

Kim YS, Stumpf WE, Sat M (1981) Anatomical distribution of estrogen target neurons in turtle brain. Brain Res 230:195–204

King JA, Millar RP (1980) Comparative aspects of vertebrate luteinizing hormone-releasing hormone structure and function in vertebrate phylogeny. Endocrinology 106:707–717

King JM, Kuchling G, Bradshaw SD (1998) Thermal environment, behaviour and body condition of wild *Pseudemydura umbrina* (Testudines: Chelidae) during late winter and early spring. Herpetologica 54:103–112

King P, Heatwole H (1997) Seasonal comparison of hemoglobins in three species of turtles. In: ASIH/HL/SSAR/AFS-ELHS/AES/GIS Joint Meetings, 26 June–2 July 1997, Seattle, pp 184

Kirkwood TBL (1981) Repair and its evolution: survival versus reproduction. In: Townsend CR, Calow P (eds) Physiological ecology: an evolutionary approach to resource use. Blackwell, Oxford, pp 165–189

Klicka J, Mahmoud IY (1977) The effects of hormones on the reproductive physiology of the painted turtle, *Chrysemys picta*. Gen Comp Endocrinol 31:407–413

Knirr M, Sachsse W, Wicker R (1997) Schildkröten. In: Köhler G (ed) Inkubation von Reptilieneiern. Herpeton Verlag Elke Köhler, Offenbach, pp 97–98

Köhler G (1997) Inkubation von Reptilieneiern. Herpeton Verlag Elke Köhler, Offenbach

Koztowski J (1992) Optimal allocation of resources to growth and reproduction – implications for age and size at maturity. Trends Ecol Evol 7:15–19

Kuchling G (1979) Zur Steuerung der Gonadenaktivität und der Winterruhe der männlichen Griechischen Landschildkröte *Testudo hermanni hermanni* Gmelin. PhD Thesis Universität Wien, Wien

Kuchling G (1981) Le cycle sexuel mâle de la tortue *Testudo hermanni hermanni* Gmelin dans une population naturelle et en captivité. Bull Soc Herpetol Fr 19:29–35

Kuchling G (1982a) Effect of temperature and photoperiod on spermatogenesis in the tortoise, *Testudo hermanni hermanni* Gmelin. Amphibia-Reptilia 2:32–341

Kuchling G (1982b) Environmental temperature, spermatogenesis and plasma testosterone concentration in the tortoise *Testudo hermanni hermanni* Gmelin. Acta Endocrinol 99 Suppl 246:29–30

Kuchling G (1986) Diurnal fluctuations of the plasma testosterone concentration in the male tortoise, *Testudo hermanni hermanni* Gmelin: the role of temperature and season. In: Assenmacher I, Boissin J (eds) Endocrine regulations as adaptive mechanisms to the environment. CNRS, Paris, pp 103–108

Kuchling G (1987) Proposal for a program to improve the captive breeding situation of the western swamp turtle *Pseudemydura umbrina*. IUCN Tortoise and Freshwater Turtle Specialist Group Newsl 2, Maitland, Florida

Kuchling G (1988a) Zur Fortpflanzung von *Pseudemydura umbrina* Siebenrock, 1901: Neue Untersuchungsmethoden für die Rettung einer vom Aussterben bedrohten Schildkrötenart (Testudines: Chelidae). Herpetozoa 1:3–11

Kuchling G (1988b) Gonadal cycles of the Western Australian long-necked turtles *Chelodina oblonga* and *Chelodina steindachneri* (Chelonia: Chelidae). Rec West Aust Mus 14:189–198

Kuchling G (1988c) Population structure, reproductive potential and increasing exploitation of the freshwater turtle *Erymnochelys madagascariensis*. Biol Conserv 43:107–113

Kuchling G (1989) Assessment of ovarian follicles and oviductal eggs by ultra-sound scanning in live freshwater turtles, *Chelodina oblonga*. Herpetologica 45:89–94

Kuchling G (1993a) Possible biennial ovarian cycle of the freshwater turtle *Erymnochelys madagascariensis*. J Herpetol 27:470–472

Kuchling G (1993b) Nesting of *Pseudemydura umbrina* (Testudines: Chelidae): the other way round. Herpetologica 49:479–487

Kuchling G (1993c) Biologie und Lebensraum von *Erymnochelys madagascariensis* (Grandidier, 1867) und Vergleich mit den anderen Wasserschildkröten Madagaskars. Salamandra 28:231–250

Kuchling G (1997a) Restoration of epidermal scute patterns during regeneration of the chelonian carapace. Chelon Conserv Biol 2:500–506

Kuchling G (1997b) Patterns of exploitation, decline, and extinction of *Erymnochelys madagascariensis*: implications for the conservation of the species. In: Abbema J Van (ed) Proceedings: Conservation, Management and Restoration of Tortoises and Turtles, 11–16 July 1993, Purchase. New York Turtle and Tortoise Society, New York, pp 113–117

Kuchling G (1997c) Conservation strategy for the natural populations of *Erymnochelys madagascariensis* at Ankarafantsika. Unpubl Report to Conservation International and Jersey Wildlife Preservation Trust. Chelonia Enterprises, Subiaco

Kuchling G, Bradshaw SD (1993) Ovarian cycle and egg production in the western swamp tortoise *Pseudemydura umbrina* (Testudines: Chelidae) in the wild and in captivity. J Zool 229:405–419

Kuchling G, DeJose JP (1989) A captive breeding operation to rescue the critically endangered western swamp turtle *Pseudemydura umbrina* from extinction. Int Zoo Yearb 28:103–109

Kuchling G, Mittermeier RA (1993) Status and exploitation of the Madagascan big-headed turtle, *Erymnochelys madagascariensis*. Chelon Conserv Biol 1:13–18

Kuchling G, Skolek-Winnisch R, Bamberg E (1981) Histochemical and biochemical investigation on the annual cycle of testis, epididymis, and plasma testosterone of the tortoise, *Testudo hermanni hermanni* Gmelin. Gen Comp Endocrinol 44:194–201

Kuchling G, DeJose JP, Burbidge AA, Bradshaw SD (1992) Beyond captive breeding: the western swamp tortoise *Pseudemydura umbrina* recovery programme. Int Zoo Yearb 31:37–41

Lance VA (1997) Sex determination in reptiles: an update. Am Zool 37:504–513

Lance VA, Callard IP (1980) Phylogenetic trends in hormonal control of gonadal steroidogenesis. In: Pang PKT, Epple A (eds) Evolution of vertebrate endocrine systems. Texas Tech Press, Lubbock, pp 167–231

Lance VA, Valenzuela N, von Hildebrand P (1992) A hormonal method to determine the sex of hatchling giant river turtles, *Podocnemis expansa*: application to endangered species research. Am Zool 32:16A

Landers JL, Garner JA, McRae WA (1980) Reproduction of gopher tortoises (*Gopherus polyphemus*) in southwestern Georgia. Herpetologica 36:353–361

Lapid R, Robinzon B (1997) Shell deformations in hatchling tortoises (*Testudo graeca*): results from X-ray radiography of mothers during pregnancy. In: Abstracts of the 3rd World Congress of Herpetology, 2–10 Aug 1997, Prague, pp 122–123

Laughran LJ, Larsen JH, Schroeder PC (1981) Ultrastructure of developing ovarian follicles and ovulation in the lizard *Anolis carolinensis* (Reptilia). Zoomorphology 98:191–208

Lee MSY (1996) Correlated progression and the origin of turtles. Nature 379:812–815

Lee MSY (1997a) Pareiasaur phylogeny and the origin of turtles. Zool J Linnean Soc 120:197–280

Lee MSY (1997b) Reptile relationships turn turtle. Nature 389:245–246

Legler JM (1958) Extra-uterine migration of ova in turtles. Herpetologica 14:49–52

Legler JM (1960) Natural history of the ornate box turtle, *Terrapene ornata ornata* Agassiz. Univ Kansas Publ Mus Nat Hist 11:527–669

Legler JM (1985) Australian chelid turtles: reproductive patterns in wide-ranging taxa. In: Grigg G, Shine R, Ehmann H (eds) Biology of Australasian frogs and reptiles. R Zool Soc New South Wales, Mosman, pp 117-123

Legler JM (1990) The genus *Pseudemys* in Mesoamerica: taxonomy, distribution, and origins. In: Gibbons JW (ed) The life history and ecology of the slider turtle. Smithsonian Institution Press, Washington, DC, pp 82-105

Legler JM (1993) Morphology and physiology of the chelonia. In: Fauna of Australia. Australian Government Publishing Service, Canberra, pp 108-119

Lever C (1990) The zoo dilemma. J Nat Hist 24:795-799

Levvy GA, Conchie J (1966) Mammalian glycosidases and their inhibition by aldonolactones. In: Neufeld EF, Ginsburg V (eds) Methods in enzymology, vol 8, Academic Press, New York, pp 571-584

Lewis J, Mahmoud IY, Klicka J (1979) Seasonal fluctuations of plasma progesterone and oestradiol-17β in the female snapping turtle, *Chelydra serpentina*. J Endocrinol 80:127-131

Licht P (1982) Endocrine patterns in the reproductive cycle of turtles. Herpetologica 38:51-61

Licht P (1983) Evolutionary divergence in the structure and function of pituitary gonadotropins of tetrapod vertebrates. Am Zool 23:673-683

Licht P (1984) Reptiles. In: Lamming GE (ed) Marshall's physiology of reproduction, vol 1: reproductive cycles of vertebrates. Churchill Livingstone, Edinburgh, pp 206-282

Licht P, Crews D (1976) Gonadotropin stimulation of in vitro progesterone production in reptilian and amphibian ovaries. Gen Comp Endocrinol 29:141-151

Licht P, Papkoff H (1985) Reevaluation of the relative activities of the pituitary glycoprotein hormones (follicle-stimulating hormone, luteinizing hormone, and thyrotrophin) from the green sea turtle, *Chelonia mydas*. Gen Comp Endocrinol 58:443-451

Licht P, Porter DA (1987) Role of gonadotropin-releasing hormone in regulation of gonadotropin secretion from amphibian and reptilian pituitaries. In: Norris DO, Jones RE (eds) Hormones and reproduction in fishes, amphibians, and reptiles. Plenum Press, New York, pp 61-85

Licht P, Wood J, Owens DW, Wood F (1979) Serum gonadotropin and steroids associated with breeding activities in the green sea turtle *Chelonia mydas*. I. Captive animals. Gen Comp Endocrinol 39:274-289

Licht P, Rainey W, Cliffton K (1980) Serum gonadotropins and steroids associated with breeding activities in the grean sea turtle, *Chelonia mydas*. II. Mating and nesting in natural populations. Gen Comp Endocrinol 40:116-122

Licht P, Breitenbach GL, Congdon JD (1985a) Seasonal cycles in testicular activity, gonadotropin, and thyroxine in the painted turtle, *Chrysemys picta*, under natural conditions. Gen Comp Endocrinol 59:130-139

Licht P, Wood JF, Wood FE (1985b) Annual and diurnal cycles in plasma testosterone and thyroxine in the male green sea turtle *Chelonia mydas*. Gen Comp Endocrinol 57:335-344

Licht P, Denver RJ, Pavgi S (1989) Temperature dependence of in vitro pituitary, testis, and thyroid secretion in a turtle, *Pseudemys scripta*. Gen Comp Endocrinol 76:274-285

Limpus CJ (1995) Global overview of the status of marine turtles: a 1995 viewpoint. In: Bjorndal KA (ed) Biology and conservation of sea turtles, revised edn. Smithsonian Institution Press, Washington London, pp 605-609

Limpus CJ, Miller JD (1993) Family Cheloniidae. In: Fauna of Australia. Australian Government Publishing Service, Canberra, pp 133-138

Limpus CJ, Nicholls N (1988) The southern oscillation regulates the annual numbers of green turtles (*Chelonia mydas*) breeding around northern Australia. Aust Wildl Res 15:157-161

Limpus CJ, Reed PC (1985) The green turtle, *Chelonia mydas*, in Queensland: a preliminary description of the population structure in a coral reef feeding ground. In: Grigg G, Shine R, Ehmann H (eds) Biology of Australasian frogs and reptiles. R Zool Soc New South Wales, Mosman, pp 47-52

Linck MH, DePari JA, Butler BO, Graham TE (1989) Nesting behavior of the turtle, *Emydoidea blandingi*, in Massachusetts. J Herpetol 23:442-444

Lindeman PV (1997) Contributions towards improvement of model fit in nonlinear regression modelling of turtle growth. Herpetologica 53:179-191

Lofts B, Boswell C (1960) Seasonal changes in the distribution of the testis lipids of the caspian terrapin *Clemmys caspica*. Proc Zool Soc Lond 136:581–592

Lofts B, Tsui HW (1977) Histological and histochemical changes in the gonads and epididymides of the male soft-shelled turtle, *Trionyx sinensis*. J Zool 181:57–68

Lohmann KJ, Lohmann CMF (1994) Acquisition of magnetic directional preference in loggerhead sea turtle hatchlings. J Exp Biol 190:1–8

Lohmann KJ, Witherington BE, Lohmann CMF, Salmon M (1997) Orientation, navigation, and natal beach homing in sea turtles. In: Lutz PL, Musick JA (eds) The biology of sea turtles. CRC Press, Boca Raton, pp 107–135

Lovich JE, McCoy CJ, Garstka WR (1990) The development and significance of melanism in the slider turtle. In: Gibbons JW (ed) The life history and ecology of the slider turtle. Smithsonian Institution Press, Washington, DC, pp 233–254

Lovich JE, Tucker AD, Kling DE, Gibbons JW, Zimmerman TD (1991) Behavior of hatchling diamondback terrapins (*Malaclemys terrapin*) released in a South Carolina salt marsh. Herpetol Rev 22:81–83

Mahapatra MS, Mahata SK, Maiti BR (1986) Circadian rhythms in serotonin, norepinephrine, and epinephrine contents of the pineal–paraphyseal complex of the soft-shelled turtle (*Lissemys punctata punctata*). Gen Comp Endocrinol 64:246–249

Mahapatra MS, Mahata SK, Maiti BR (1989) Effect of ambient temperature on serotonin, norepinephrine, and epinephrine contents in the pineal–paraphyseal complex of the soft-shelled turtle (*Lissemys punctata punctata*). Gen Comp Endocrinol 74:215–220

Mahapatra MS, Mahata SK, Maiti BR (1991) Effect of stress on serotonin, norepinephrine, epinephrine and corticosterone contents in the soft-shelled turtle. Clin Exp Pharmacol Physiol 18:719–724

Mahata M, Mahata SK (1992) Effect of steroid hormones on serotonin, norepinephrine and epinephrine contents in the pineal–paraphyseal complex of the soft-shelled turtle (*Lissemys punctata punctata*). J Comp Physiol B 162:520–525

Mahata-Mahapatra M, Mahata SK (1991) Circannual pineal rhythms in the soft-shelled turtle (*Lissemys punctata punctata*). J Interdiscip Cycle Res 23:9–16

Mahmoud IY, Licht P (1997) Seasonal changes in gonadal activity and the effects of stress on reproductive hormones in the common snapping turtle, *Chelydra serpentina*. Gen Comp Endocrinol 107:359–372

Mahmoud IY, Hess GL, Klicka J (1973) Normal embryonic stages of the western painted turtle, *Chrysemys picta belli*. J Morphol 141:269–279

Mahmoud IY, Cyrus RV, Bennett TM, Woller MJ, Montag DM (1985) Ultrastructural changes in testes of the snapping turtle *Chelydra serpentina* in relation to plasma testosterone, Δ^5-3β-hydroxysteroid dehydrogenase, and cholesterol. Gen Comp Endocrinol 57:454–464

Mahmoud IY, Cyrus RV, McAsey ME, Cady C, Woller MJ (1988) The role of arginine vasotocin and prostaglandin $F_{2\alpha}$ on oviposition and luteolysis in the common snapping turtle, *Chelydra serpentina*. Gen Comp Endocrinol 69:56–64

Mahmoud IY, Guillette LJ Jr, McAsey ME, Cady C (1989) Stress-induced changes in serum testosterone, estradiol-17β and progesterone in the turtle, *Chelydra serpentina*. Comp Biochem Physiol A 93:433–427

Mallinson JJC (1991) Partnerships for conservation between zoos, local governments and non-governmental organizations. Symp Zool Soc Lond 62:57–74

Manton ML (1979) Olfaction and behavior. In: Harless M, Morlock H (eds) Turtles perspectives and research. John Wiley, New York, pp 289–301

Mazzi V, Vellano C (1987) Prolactin and reproduction. In: Norris DO, Jones RE (eds) Hormones and reproduction in fishes, amphibians, and reptiles. Plenum Press, New York, pp 87–115

McIntyre S, Barrett GW, Kitching RL, Recher HF (1992) Species triage – seeing beyond wounded rhinos. Conserv Biol 6:604–606

McKeown S, Meier DE, Juvik JO (1990) The management and breeding of the Asian forest tortoise (*Manouria emys*) in captivity. In: Beaman KR, Caporaso F, McKeown S, Graff M (eds) Proceedings of the first international symposium on turtles and tortoises: conservation and captive husbandry. California Turtle and Tortoise Club, Van Nuys, pp 138–159

McKnight CM, Gutzke WHN (1993) Effects of the embryonic environment and of hatchling housing conditions on growth of young snapping turtles (*Chelydra serpentina*). Copeia 1993:475–482

McPherson RJ, Marion KR (1981a) The reproductive biology of female *Sternotherus odoratus* in an Alabama population. J Herpetol 15:389–396

McPherson RJ, Marion KR (1981b) Seasonal testicular cycle of the stinkpot turtle (*Sternotherus odoratus*) in central Alabama. Herpetologica 37:33–40

McPherson RJ, Marion KR (1982) Seasonal changes of total lipids in the turtle *Sternotherus odoratus*. Comp Biochem Physiol 71A:93–98

McPherson RJ, Boots LR, MacGregor R III, Marion KR (1982) Plasma steroids associated with seasonal reproductive changes in a multiclutched freshwater turtle, *Sternotherus odoratus*. Gen Comp Endocrinol 48:440–451

Meissl H, Ueck M (1980) Extraocular photoreception of the pineal gland of the aquatic turtle *Pseudemys scripta elegans*. J Comp Physiol 140:173–179

Mendonça MT (1987a) Photothermal effects on the ovarian cycle on the musk turtle, *Sternotherus odoratus*. Herpetologica 43:82–90

Mendonça MT (1987b) Timing of reproductive behaviour in male musk turtles, *Sternotherus odoratus*: effects of photoperiod, temperature and testosterone. Anim Behav 35:1002–1014

Mendonça MT, Licht P (1986a) Seasonal cycles in gonadal activity and plasma gonadotropin in the musk turtle, *Sternotherus odoratus*. Gen Comp Endocrinol 62:459–469

Mendonça MT, Licht P (1986b) Photothermal effects on the testicular cycle in the musk turtle, *Sternotherus odoratus*. J Exp Zool 239:117–130

Merchant-Larios H, Ruiz-Ramirez S, Moreno-Mendoza N, Marmolejo-Valencia A (1997) Correlation among thermosensitive period, estradiol response, and gonadal differentiation in the sea turtle *Lepidochelys olivacea*. Gen Comp Endocrinol 107:373–385

Mesner PW, Mahmoud IY, Cyrus RV (1993) Seasonal testosterone levels in Leydig and Sertoli cells of the snapping turtle (*Chelydra serpentina*) in natural populations. J Exp Zool 266:266–276

Meylan A (1995) Sea turtle migration – evidence from tag returns. In: Bjorndal KA (ed) Biology and conservation of sea turtles, revised edn. Smithsonian Institution Press, Washington, pp 91–100

Miller JD (1985) Embryology of marine turtles. In: Gans C, Billet F, Maderson P (eds) Biology of the Reptilia, vol 14. John Wiley, New York, pp 269–328

Miller JD (1997) Reproduction in sea turtles. In: Lutz PL, Musick JA (eds) The biology of sea turtles. CRC Press, Boca Raton, pp 51–81

Miller K (1987) Hydric conditions during incubation influence locomotor performance of hatchling snapping turtles. J Exp Biol 127:401–412

Mittermeier RA (1978) South America's river turtles: saving them by use. Oryx 14:222–230

Moll D (1994) The ecology of sea beach nesting in slider turtles (*Trachemys scripta venusta*) from Caribbean Costa Rica. Chelon Conserv Biol 1:107–116

Moll D (1995) Conservation and management of river turtles: a review of methodology and techniques. In: SOPTOM (ed) Proceedings, International Congress of Chelonian Conservation, 6–10 July 1995, Gonfaron. Editions SOPTOM, Gonfaron, pp 290–294

Moll D, Moll EO (1990) The slider turtle in the neotropics: adaptation of a temperate species to a tropical environment. In: Gibbons JW (ed) The life history and ecology of the slider turtle. Smithsonian Institution Press, Washington, DC, pp 152–161

Moll EO (1979) Reproductive cycles and adaptations. In: Harless M, Morlock H (eds) Turtles perspectives and research. John Wiley, New York, pp 305–331

Moll EO (1991) India's freshwater turtle resource with recommendations for management. In: Daniel JC, Serrao JS (eds) Conservation in developing countries: problems and prospects. Bombay Nat Hist Soc and Oxford Univ Press, Bombay, pp 501–515

Moll EO, Matson EK, Krehbiel EB (1981) Sexual and seasonal dichromatism in the Asian river turtle *Callagur borneoensis*. Herpetologica 37:181–194

Moore MC, Lindzey J (1992) The physiological basis of sexual behavior in male reptiles. In: Gans C, Crews D (eds) Biology of Reptilia, vol. 18. Univ Chicago Press, Chicago, pp 70–113

Morales MH, Sanchez EJ (1996) Changes in vitellogenin expression during captivity-induced stress in a tropical anole. Gen Comp Endocrinol 103:209–219

Mortimer JA (1995) Headstarting as a management tool. In: Bjorndal KA (ed) Biology and conservation of sea turtles, revised edn. Smithsonian Institution Press, Washington, pp 613–615

Moskovits DK (1988) Sexual dimorphism and population estimates of the two Amazonian tortoises (*Geochelone carbonaria* and *G. denticulata*) in northwestern Brazil. Herpetologica 44:209–217

Mrosovsky N (1997) IUCN's credibility critically endangered. Nature 389:436

Mrosovsky N, Pieau C (1991) Transitional range of temperature, pivotal temperatures and thermosensitive stages for sex determination in reptiles. Amphibia-Reptilia 12:169–179

Mulaa FJ, Aboderin AA (1992) Two phosphoglycoproteins (phosvitins) from *Kinixys erosa* oocytes. Comp Biochem Physiol 103B:1025–1031

Nagahama Y (1987) Endocrine control of oocyte maturation. In: Norris DO, Jones RE (eds) Hormones and reproduction in fishes, amphibians, and reptiles. Plenum Press, New York, pp 171–202

Nagy K, Medica PA (1986) Physiological ecology of desert tortoises in southern Nevada. Herpetologica 42:73–92

Nagy KA, Morafka DJ, Yates RA (1997) Young desert tortoise survival: energy, water, and food requirements in the field. Chelon Conserv Biol 2:396–404

Naulleau G, Bonnet X (1996) Body condition threshold for breeding in a viviparous snake. Oecologia 107:301–306

Nieuwolt-Dacanay PM (1997) Reproduction in the western box turtle, *Terrapene ornata luteola*. Copeia 1997:819–826

Noegel RP, Moss GA (1989) Breeding the Galapagos tortoise *Geochelone elephantopus* at the Life Fellowship Bird Sanctuary, Seffner. Int Zoo Yearb 28:78–83

Northcote TG (1984) Mechanisms of fish migration in rivers. In: McCleave JD, Arnold GP, Dodson JJ, Neil WH (eds) Mechanisms of migration in fishes. Proc NATO Adv Res mechanisms of migration in fishes. Plenum Press, New York, pp 317–355

Nozaki M, Tsukahara T, Kobayashi H (1984) Neuronal systems producing LHRH in vertebrates. In: Ochiai K, Arai V, Shioda T, Takahashi M (eds) Endocrine correlates of reproduction. Japan Sci Soc Press, Tokyo, pp 3–27

Obbard ME, Brooks RJ (1980) Nesting migrations of the snapping turtle (*Chelydra serpentina*). Herpetologica 36:158–162

Obst FJ (1985) Die Welt der Schildkröten. Albert Müller Verlag, Rüschlikon-Zürich

O'Malley BW, McGuire WL, Kohler PO, Korenman SG (1969) Studies on the mechanism of steroid hormone regulation of synthesis of specific protein. In: Astwood EB (ed) Recent progress in hormone research, vol 25. Academic Press, New York, pp 105–160

Owens DW (1980) The comparative reproductive physiology of sea turtles. Am Zool 20:549–563

Owens DW (1995) The role of reproductive physiology in the conservation of sea turtles. In: Bjorndal KA (ed) Biology and conservation of sea turtles, revised edn. Smithsonian Institution Press, Washington, pp 39–44

Owens DW (1997) Hormones in the life history of sea turtles. In: Lutz PL, Musick JA (eds) The biology of sea turtles. CRC Press, Boca Raton, pp 315–341

Owens DW, Morris YA (1985) The comparaive endocrinology of sea turtles. Copeia 1985:723–735

Owens DW, Ruiz GJ (1980) New methods of obtaining blood and cerebrospinal fluid from marine turtles. Herpetologica 36:17–20

Owens DW, Gern WA, Ralph CL (1980) Melatonin in the blood and cerebrospinal fluid of the green sea turtle (*Chelonia mydas*). Gen Comp Endocrinol 40:180–187

Owens DW, Grassman MA, Hendrickson JR (1982) The imprinting hypothesis and sea turtle reproduction. Herpetologica 38:124–135

Packard GC, Packard MJ (1988) The physiological ecology of reptilian eggs and embryos. In: Gans C, Huey RB (eds) Biology of the Reptilia, vol 16. Alan R Liss, New York, pp 423–605

Packard GC, Packard MJ (1997) Overwintering strategies of neonatal turtles. In: Abstracts of the 3rd World Congress of Herpetology, 2–10 Aug 1997, Prague, pp 157

Packard GC, Fasano SL, Attaway MB, Lohmiller LD, Lynch TL (1997a) Thermal environment for overwintering hatchlings of the painted turtle (*Chrysemys picta*). Can J Zool 75:401–406

Packard GC, Lang JW, Lohmiller LD, Packard MJ (1997b) Cold tolerance in hatchling painted turtles (*Chrysemys picta*): supercooling or tolerance for freezing? Physiol Zool 70:670–678

Packard MJ, Packard GC, Boardman TJ (1982) Structure of eggshells and water relations of reptilian eggs. Herpetologica 38:136–155

Palmer BD, Guillette LJ Jr (1988) Histology and functional morphology of the female reproductive tract of the tortoise *Gopherus polyphemus*. Am J Anat 183:200–211

Palmer BD, Guillette LJ Jr (1990) Morphological changes in the oviductal endometrium during the reproductive cycle of the tortoise, *Gopherus polyphemus*. J Morphol 204:323–333

Parmenter CJ (1976) The natural history of the Australian freshwater turtle *Chelodina longicollis* Shaw (Testudinata, Chelidae). PhD Thesis University of New England, Armidale

Parmenter CJ (1985) Reproduction and survivorship of *Chelodina longicollis* (Testudinata: Chelidae). In: Grigg G, Shine R, Ehmann H (eds) Biology of Australasian frogs and reptiles. R Zool Soc New South Wales, Mosman, pp 53–61

Parmenter RR, Avery HW (1990) The feeding ecology of the slider turtle. In: Gibbons JW (ed) The life history and ecology of the slider turtle. Smithsonian Institution Press, Washington, DC, pp 257–266

Penninck DG, Stewart JS, Paul-Murphy J, Pion P (1991) Ultrasonography of the Californian desert tortoise (*Xerobates agassizi*): anatomy and application. Vet Radiol 32:112–116

Perez LE, Williams D, Callard IP (1992) Putative apolipoprotein B-100 in the freshwater turtle *Chrysemys picta*: effects of estrogen and progesterone. Comp Biochem Physiol 103B:707–713

Perezfigares JM, Mancera JM, Rodriguez EM, Nualart F, Fernandezllebrez P (1995) Presence of an oxytocin-like peptide in the hypothalamus and neurohypophysis of a turtle (*Mauremys caspica*) and a snake (*Natrix maura*). Cell Tissue Res 279:75–84

Phillips CA, Dimmick WW, Carr JL (1996) Conservation genetics of the common snapping turtle (*Chelydra serpentina*). Conserv Biol 10:397–405

Pianka ER (1970) On r- and K-selection. Am Nat 104:592–597

Pieau C (1971) Sur la proportion sexuelle chez les embryons de deux Chéloniens (*Testudo graeca* L. et *Emys orbicularis* L.) issus d'oeufs incubés artificiellement. C R Seanc Acad Sci Paris 272D:3071–3074

Pieau C (1972) Effets de la température sur le dévelopement des glandes génitales chez les embryons de deux Chéloniens, *Emys orbicularis* L. et *Testudo graeca* L. C R Acad Sci Paris 274D:719–722

Pieau C (1996) Temperature variation and sex determination in reptiles. Bioessays 18:19–26

Platz JE, Conlon JM (1997) And turn back again. Nature 389:246

Plotkin PT, Rostal DC, Byles RA, Owens DW (1997) Reproductive and developmental synchrony in female *Lepidochelys olivacea*. J Herpetol 31:17–22

Polisar J (1996) Reproductive biology of a flood-season nesting freshwater turtle of the northern neotropics: *Dermatemys mawii* in Belize. Chelon Conserv Biol 2:13–25

Pritchard PCH (1979a) Encyclopedia of turtles. T.F.H. Publications, Neptune

Pritchard PCH (1979b) Taxonomy, evolution, and zoogeography. In: Harless M, Morlock H (eds) Turtles perspectives and research. John Wiley, New York, pp 1–42

Pritchard PCH (1995) Introduction: the turtle campaign. In: SOPTOM (ed) Proceedings, International Congress of Chelonian Conservation, 6–10 July 1995, Gonfaron. Editions SOPTOM, Gonfaron, pp 1–3

Pritchard PCH (1996) The Galapagos tortoises' nomenclatural and survival status. Chelon Res Monogr 1:1–85

Pritchard PCH (1997a) Evolution, phylogeny, and current status. In: Lutz PL, Musick JA (eds) The biology of sea turtles. CRC Press, Boca Raton, pp 1–28

Pritchard PCH (1997b) Conservation strategies – an overview: implications for management. In: Abbema J van (ed) Proceedings: conservation, management and restoraton of tortoises and turtles, 11–16 July 1993, Purchase. New York Turtle and Tortoise Society, New York, pp 467–471

Pritchard PCH, Trebbau P (1984) The turtles of Venezuela. Contrib Herpetol 2:1–403

Quinn NWS, Tate DP (1991) Seasonal movements and habitat of wood turtles (*Clemmys insculpta*) in Algonquin Park, Canada. J Herpetol 25:217–220

Rao RJ (1995) Conservation status of Indian chelonians. In: SOPTOM (ed) Proceedings, International Congress of Chelonian Conservation, 6–10 July 1995, Gonfaron. Editions SOPTOM, Gonfaron, pp 33–37

Ravet V, Depeiges A, Morel F, Dufaure JP (1991) Synthesis and post-translational modifications of an epididymal androgen dependent protein family. Gen Comp Endocrinol 84:104–114

Reichman OJ (1984) Evolution of regeneration capabilities. Am Nat 123:752–763

Rhen T, Lang JW (1995) Phenotypic plasticity for growth in the common snapping turtle – effects of incubation temperature, clutch, and their interaction. Am Nat 146:726–747

Ricceri G (1953) Ricerche biochimiche sul letargo dei cheloni. Boll Acad Gioenia 4:370–390

Risley PL (1933a) Contributions on the development of the reproductive system in the musk turtle, *Sternotherus odoratus* (Latreille). I. The embryonic origin and migration of the primordial germ cells. Z Zellforsch 18:459–492

Risley PL (1933b) Contributions on the development of the reproductive system in the musk turtle, *Sternotherus odoratus* (Latreille). II. Gonadogenesis and sex differentiation. Z Zellforsch 18:493–541

Risley P (1937) A preliminary study of sex development in turtle embryos following administration of testosterone. Anat Rec 70:103

Risley P (1939) Effects of gonadotropic and sex hormones on the urigenital system of the juvenile diamond-back terrapins. Anat Rec 75(Suppl):104

Robeck TR, Rostal DC, Burchfield PM, Owens DW, Kraemer DC (1990) Ultrasound imaging of reproductive organs and eggs in Galapagos tortoises, *Geochelone elephantopus* spp. Zoo Biol 9:349–359

Rogner M (1995) Schildkröten 1 Chelydridae Dermatemydidae Emydidae. Heidi Rogner-Verlag, Hürtgenwald

Rogner M (1996) Schildkröten 2. Heidi Rogner-Verlag, Hürtgenwald

Rohr W (1970) Die Bedeutung des Wärmefaktors für Fortpflanzungsperiodik und Eiablageverhalten südeuropäischer Landschildkröten im Terrarium. Salamandra 6:99–103

Roosenburg WM, Kelley KC (1996) The effect of egg size and incubation temperature on growth in the turtle, *Malaclemys terrapin*. J Herpetol 30:198–204

Rose FL (1986) Carapace regeneration in *Terrapene* (Chelonia: Testudinidae). Southwest Nat 31:131–134

Rose FL, Simpson TR, Manning RW (1996) Measured and predicted egg volume of *Pseudemys texana* with comments on turtle egg shape. J Herpetol 30:433–435

Rostal DC, Robeck TR, Owen DW, Kraemer DC (1990) Ultrasound imaging of ovaries and eggs in Kemp's ridley sea turtles (*Lepidochelys kempi*). J Zoo Wildl Med 21:27–35

Rostal DC, Lance VA, Grumbles JS, Alberts AC (1994) Seasonal reproductive cycles of the desert tortoise (*Gopherus agassizii*) in the eastern Mojave Desert. Herpetol Monogr 8:72–82

Rostal DC, Paladino FV, Patterson RM, Spotila JR (1996) Reproductive physiology of nesting leatherback turtles (*Dermochelys coriacea*) at Las Baulas national park, Costa Rica. Chelon Conserv Biol 2:230–236

Rowe JW (1994) Reproductive variation and egg size–clutch size trade-off within and among populations of painted turtles (*Chrysemys picta bellii*). Oecologia 99:35–44

Rowe JW, Holy L, Ballinger RE, Stanley-Samuelson D (1995) Lipid provisioning of turtle eggs and hatchlings: total lipid, phospholipid, triacylglycerol and triacylglycerol fatty acids. Comp Biochem Physiol 112B:323–330

Ryan KM, Spotila JR, Standora EA (1990) Incubation temperature and post-hatching growth and performance in snapping turtles. Am Zool 30:112A

Saint Girons H (1963) Spermatogenese et évolution cyclique des caractères sexuels secondaires chez les squamata. Ann Sci Nat Zool 5:461–478

Saint Girons H (1975) Sperm survival and transport in the female genital tract of reptiles. In: Hafez ESE, Thibault CG (eds) The biology of spermatozoa. S Karger, Basel, pp 105–113

Sakamoto W, Bando T, Arai N, Baba N (1997) Migration paths of the adult female and male loggerhead turtles *Caretta caretta* determined through satellite telemetry. Fish Sci 63:547–552

Sarkar S, Sarkar NK, Maiti BR (1995) Histological and functional changes of oviductal endometrium during seasonal reproductive cycle of the soft-shelled turtle, *Lissemys punctata punctata*. J Morphol 224:1–14

Sarkar S, Sarkar NK, Das P, Maiti BR (1996a) Photothermal effects on ovarian growth and function in the soft-shelled turtle *Lissemys punctata punctata*. J Exp Zool 274:41–55

Sarkar S, Sarkar NK, Maiti BR (1996b) Seasonal pattern of ovarian growth and interrelated changes in plasma steroid levels, vitellogenesis, and oviductal function in the adult female soft-shelled turtle *Lissemys punctata punctata*. Can J Zool 74:303–311

Schaefer I (1986) Haltung und Nachzucht der Fransenschildkröte *Chelus fimbriatus* (Schneider, 1783) (Testudines: Chelidae). Salamandra 22:229–241

Schildger BJ, Baumgartner R, Häfeli W, Rübel A, Isenbügel E (1993) Narkose und Immobilisation bei Reptilien. Tierärztl Prax 21:361–376

Scott NJ, Rathbun GB, Murphey TG, Harker MB (1997) Reproductive parameters of pacific pond turtles, *Clemmys marmorata*, in streams on the coast of central California. In: ASIH/HL/SSAR/AFS-ELHS/AES/GIS Joint Meetings, 26 June–2 July 1997, Seattle, pp 264

Seidel ME, Fritz U (1997) Courtship behavior provides additional evidence for a monophyletic *Pseudemys*, and comments on mesoAmerican *Trachemys* (Testudines: Emydidae). Herpetol Rev 28:70–72

Sen M, Maiti BR (1990) Histomorphological changes of the ovary and oviduct during sexual maturity in the soft-shelled turtle, *Lissemys punctata punctata*. Zool Anz 225:391–395

Seymour RS, Kennett R, Christian K (1997) Osmotic balance in the eggs of the turtle *Chelodina rugosa* during developmental arrest under water. Physiol Zool 70:301–306

Shaffer HB, Meylan P, McKnight ML (1997) Tests of turtle phylogeny: molecular, morphological, and paleontological approaches. Syst Biol 46:235–268

Shine R (1985) The evolution of reptilian viviparity: an ecological analysis. In: Gans C, Billett F (eds) Biology of the Reptilia, vol 15. John Wiley, New York, pp 605–694

Shine R (1988) Parental care in reptiles. In: Gans C, Huey RB (eds) Biology of the Reptilia, vol 16. Alan R Liss, New York, pp 275–329

Shine R, Iverson JB (1995) Patterns of survival, growth and maturation in turtles. Oikos 72:343–348

Silva AMR, Morales GS, Wassermann GF (1984) Seasonal variations of testicular morphology and plasma levels of testosterone in the turtle *Chrysemys dorbigni*. Comp Biochem Physiol 78A:153–157

Singh DP (1974) Analysis of environmental factors regulating the gonadal cycle in a tropical pond turtle, *Lissemys p. granosa* (Schoepff.). Experientia 30:967–968

Singh DP (1977) Annual sexual rhythm in relation to environmental factors in a tropical pond turtle, *Lissemys punctata granosa*. Herpetologica 33:190–194

Smith HM (1958) Total regeneration of the carapace in a box turtle. Turtox News 36:234–238

Smith JM (1991) The evolution of reproductive strategies: a commentary. Philos Trans R Soc Lond B 332:103–104

SOPTOM (1995) Proceedings, International Congress of Chelonian Conservation, 6–10 July 1995, Gonfaron. Editions SOPTOM, Gonfaron

Soulé ME (1987) Viable populations for conservation. Cambridge University Press, Cambridge

Souza RR De, Vogt RC (1994) Incubation temperature influences sex and hatchling size in the neotropical turtle *Podocnemis unifilis*. J Herpetol 28:453–464

Spence T, Fairfax R, Loach I (1979) The Western Australian swamp tortoise *Pseudemydura umbrina* in captivity. Int Zoo Yearb 19:58–60

Spotila JR, Zimmerman LC, Binckley CA, Grumbles JS, Rostal DC, List A Jr, Beyer EC, Phillips KM, Kemp SJ (1994) Effects of incubation conditions on sex determination, hatching success, and growth of hatchling desert tortoises, *Gopherus agassizii*. Herpetol Monogr 8:103–116

Sprando RL, Russell LD (1988) Spermiogenesis in the red-eared turtle (*Pseudemys scripta*) and the domestic fowl (*Gallus domesticus*): a study of cytoplasmic events including cell volume changes and cytoplasmatic elimination. J Morphol 198:95–118

Stancyk SE (1995) Non-human predators of sea turtles and their control. In: Bjorndal KA (ed) Biology and conservation of sea turtles, revised edn. Smithsonian Institution Press, Washington, pp 139–152

Stearns SC (1992) The evolution of life histories. Oxford Univ Press, Oxford

Stephens GA, Creekmore JS (1983) Blood collection by cardiac puncture in conscious turtles. Copeia 1983:522–523

Stettner A (1996) Warum ist der so lästig? Paarungswütige Schildkrötenmänner. Emys 3:25–27

Storey KB, Storey JM, Brooks SPJ, Churchill TA, Brooks RJ (1988) Hatchling turtles survive freezing during winter hibernation. Proc Natl Acad Sci USA 85:8350–8354

Stubbs D, Swingland IR, Hailey A (1985) The ecology of the Mediterranean tortoise *Testudo hermanni* in northern Greece (the effects of a catastrophe on population structure and density). Biol Conserv 31:125–152

Suzuki H, Yamamoto T, Kikuyama S, Oguchi A, Uemura H (1997) Distribution of endothelin 3-like immunoreactivity in bullfrog and soft-shelled turtle pituitaries. In: Kawashima S, Kikuyama S (eds) Advances in comparative endocrinology. Proc XIIIth Int Congress Comp Endocrinol, 16–21 Nov 1997. Monduzzi Editore, Bologna, pp 789–793

Swingland IR (1989a) *Geochelone elephantopus* Galapagos giant tortoise. In: Swingland IR, Klemens W (eds) The conservation biology of tortoises. Occasional Papers of the IUCN Species Survival Commission No 5, IUCN, Gland, pp 24–28

Swingland IR (1989b) *Geochelone gigantea* Aldabra giant tortoise. In: Swingland IR, Klemens W (eds) The conservation biology of tortoises. Occasional Papers of the IUCN Species Survival Commission No 5, IUCN, Gland, pp 105–110

Swingland IR, Coe MJ (1978) The natural regulation of giant tortoise populations on Aldabra Atoll: reproduction. J Zool 186:285–309

Swingland IR, Lessells CM (1979) The natural regulation of giant tortoise populations on Aldabra Atoll. Movement polymorphism, reproductive success and mortality. J Anim Ecol 48:639–654

Swingland IR, North PM, Dennis A, Parker MJ (1989) Movement patterns and morphometrics in giant tortoises. J Anim Ecol 58:971–985

Thinès G (1968) Activity regulation in the tortoise *Testudo hermanni* Gmelin. Psychol Belg 8:131–138

Thomas RB, Beckman DW, Thompson K, Buhlmann KA, Gibbons JW, Moll DL (1997) Estimation of age for *Trachemys scripta* and *Deirochelys reticularia* by counting annual growth layers in claws. Copeia 1997:842–845

Thompson MB (1988) Influence of incubation temperature and water potential on sex determination in *Emydura macquarii* (Testudines: Pleurodira). Herpetologica 44:86–90

Thomson S, Adams M, Seddon J, Georges A (1997) The Western Australian turtle *Chelodina oblonga* (Testudines: Chelidae) and its phylogenetic placement within the genus *Chelodina*. In: ASIH/HL/SSAR/AFS-ELHS/AES/GIS Joint Meetings, 26 June–2 July 1997, Seattle, pp 290

Thorbjarnarson JB, Perez N, Escalona T (1993) Nesting of *Podocnemis unifilis* in the Capanaparo river, Venezuela. J Herpetol 27:344–347

Tinkle DW, Congdon JD, Rosen PC (1981) Nesting frequency and success: implications for the demography of painted turtles. Ecology 62:1426–1432

Townsend CH (1925) The Galapagos tortoises in their relation to the whaling industry. Zoologica 4:55–135

Tronc E, Vuillemin S (1973) Contribution á l'étude ostéologique de *Erymnochelys madagascariensis* Grandidier, 1867. Bull Acad Malg 51:189–224

Tuberville TD, Gibbons JW, Greene JL (1996) Invasion of new aquatic habitats by male freshwater turtles. Copeia 1996:713–715

Underwood H (1992) Endogenous rhythms. In: Gans C, Crews D (eds) Biology of Reptilia, vol 18. Univ Chicago Press, Chicago, pp 229–297

Valenzuela N, Botero R, Martínez E (1997) Field study of sex determination in *Podocnemis expansa* from Colombian Amazonia. Herpetologica 53:390–398

Van Denburgh J (1914) The gigantic land tortoises of the Galapagos Islands. Proc Calif Acad Sci 4th Ser 2:203–374

Vanha-Perttula T (1978) Spermatogenesis and hydrolytic enzymes – a review. Ann Biol Anim Biochem Biophys 18:633–644

Van Nassauw L, Callebaut M (1987) Immunohistochemical localization of desmin in the ovary of the turtle, *Pseudemys scripta elegans*. Med Sci Res 15:361–362

Vivien-Roels B, Arendt J (1981) Environmental control of pineal and gonadal function in reptiles: preliminary results on the relative roles of photoperiod and temperature. Les Colloques de l'INRA 6:273–288

Vivien-Roels B, Arendt J, Bradtke J (1979) Circadian and circannual fluctuations of pineal indoleamines (serotonin and melatonin) in *Testudo hermanni* Gmelin (Reptilia, Chelonia). Gen Comp Endocrinol 37:197–210

Vivien-Roels B, Pévet P, Claustrat B (1988) Pineal and circulating melatonin rhythms in the box turtle, *Terrapene carolina triunguis*: effect of photoperiod, light pulse, and environmental temperature. Gen Comp Endocrinol 69:163–173

Vogt RC (1995) Brazilian freshwater turtles, to eat or not to eat? Save them by eating them? In: SOPTOM (ed) Proceedings, International Congress of Chelonian Conservation, 6–10 July 1995, Gonfaron. Editions SOPTOM, Gonfaron, pp 151–154

Vogt RC (1997a) Freshwater turtle nesting and reproduction in the neotropics. In: ASIH/HL/SSAR/AFS-ELHS/AES/GIS Joint Meetings, 26 June–2 July 1997, Seattle, pp 301

Vogt RC (1997b) Sexual maturity in female turtles: is it age or size? In: ASIH/HL/SSAR/AFS-ELHS/AES/GIS Joint Meetings, 26 June–2 July 1997, Seattle, pp 301

Walker TA, Parmenter CJ (1990) Absence of a pelagic phase in the life cycle of the flatback turtle, *Natator depressa* (Garman). J Biogeogr 17:275–278

Walker WF (1979) Locomotion. In: Harless M, Morlock H (eds) Turtles perspectives and research. John Wiley, New York, pp 435–454

Webb GJW (1997a) Sustainable use of wildlife. Aust Biol 10:3–11

Webb GJW (1997b) Crocodiles. Aust Biol 10:31–39

Webb GJW, Cooper-Preston H (1989) Effects of incubation temperature on crocodiles and the evolution of reptilian oviparity. Am Zool 29:953–971

Weindl A, Kuchling G (1982) Immunohistochemistry of somatostatin in the central nervous system of the tortoise *Testudo hermanni* Gmelin. Verh Dtsch Zool Ges 1982:209

Weindl A, Kuchling G, Triepel J, Reinecke M (1983a) Immunohistochemical localization of substance P in the brain and spinal cord of the tortoise *Testudo hermanni* Gmelin. In: Skrabanek P, Powell D (eds) Substance P. Boole Press, Dublin, pp 265–266

Weindl A, Kuchling G, Wetzstein R (1983b) The distribution of neurohypophyseal peptides in the central nervous system of the tortoise *Testudo hermanni* Gmelin. Acta Endocrinol 102 Suppl 253:67–68

Weindl A, Triepel J, Kuchling G (1984) Somatostatin in the brain of the turtle *Testudo hermanni* Gmelin: an immunohistochemical mapping study. Peptides 5 Suppl 1:91–100

Weismann A (1885) Die Kontinuität des Keimplasmas als Grundlage einer Theorie der Vererbung. Gustav Fischer, Jena

White JB, Murphy GG (1973) The reproductive cycle and sexual dimorphism of the common snapping turtle, *Chelydra serpentina serpentina*. Herpetologica 29:240–246

Whittier JM, Crews D (1987) Seasonal reproduction: patterns and control. In: Norris DO, Jones RE (eds) Hormones and reproduction in fishes, amphibians, and reptiles. Plenum Press, New York, pp 385–409

Whittier JM, Tokarz RR (1992) Physiological regulation of sexual behavior in female reptiles. In: Gans C, Crews D (eds) Biology of Reptilia, vol 18. Univ Chicago Press, Chicago, pp 24–69

Whittier JM, Corrie F, Limpus C (1997) Plasma steroid profiles in nesting loggerhead turtles (*Caretta caretta*) in Queensland, Australia: relationship to nesting episode and season. Gen Comp Endocrinol 106:39–47

Wibbels T, MacKenzie DS, Owens DW, Amoss S, Limpus CJ (1986) Aspects of thyroid physiology in loggerhead sea turtles, *Caretta caretta*. Am Zool 26:564

Wibbels T, Owens DW, Limpus C, Reed P, Amoss M (1990) Seasonal changes in gonadal steroid concentrations associated with migration, mating, and nesting in loggerhead sea turtles. Gen Comp Endocrinol 79:154–164

Wibbels T, Owens DW, Licht P, Limpus C, Reed P, Amoss M (1992) Serum gonadotropins and gonadal steroids associated with ovulation and egg production in sea turtles. Gen Comp Endocrinol 87:71–78

Wicker R (1997) Krötenkopfschildkröten. In: Köhler G (ed) Inkubation von Reptilieneiern. Herpeton Verlag Elke Köhler, Offenbach, 103 pp

Wilbur HM (1975) The evolutionary and mathematical demography of the turtle *Chrysemys picta*. Ecology 56:64–77

Wilbur HM, Morin PJ (1988) Life history evolution in turtles. In: Gans C, Huey RB (eds) Biology of the Reptilia, vol 16. Alan R Liss, New York, pp 387–439

Wilkinson M, Thorley J, Benton MJ (1997) Uncertain turtle relationships. Nature 387:466

Williams GC (1957) Pleiotropy, natural selection and the evolution of senescence. Evolution 11:398–411

Williams GC (1966) Adaptation and natural selection. Princeton Univ Press, Princeton

Williams GC (1992) Natural selection: domains, levels, and challenges. Oxford University Press, New York

Witherington BE (1995) Hatchling orientation. In: Bjorndal KA (ed) Biology and conservation of sea turtles, revised edn. Smithsonian Institution Press, Washington, pp 577–578

Wood F (1991) Turtle culture. In: Nash CE (ed) Production of aquatic animals: crustaceans, molluscs, amphibians and reptiles. Elsevier, Amsterdam, pp 225–234

Wood F, Wood J (1990) Successful production of captive F2 generation of the green sea turtle. Mar Turtle Newsl 50:3–4

Woodbury AM, Hardy R (1948) Studies of the desert tortoise, *Gopherus agassizii*. Ecol Monogr 18:145–200

Wyneken J, Salmon M, Lohmann KJ (1990) Orientaion by hatchling loggerhead sea turtles *Caretta caretta* L. in a wave tank. J Exp Mar Biol Ecol 139:43–50

Xavier F (1987) Functional morphology and regulation of the corpus luteum. In: Norris DO, Jones RE (eds) Hormones and reproduction in fishes, amphibians, and reptiles. Plenum Press, New York, pp 241–282

Yasukawa Y, Ota H, Iverson JB (1996) Geographic variation and sexual size dimorphism in *Mauremys mutica* (Cantor, 1842) (Reptilia: Bataguridaw), with description of a new subspecies from the southern Ryukyus, Japan. Zool Sci 13:303–317

Yntema CL (1968) A series of stages in the embryonic development of *Chelydra serpentina*. J Morphol 125:219–252

Yu JYL (1997) Functional evolution of pituitary glycoprotein hormones and receptors in vertebrates: hormone–receptor interaction. In: Kawashima S, Kikuyama S (eds) Advances in comparative endocrinology. Proc XIIIth Int Congress Comp Endocrinol, 16–21 Nov 1997. Monduzzi Editore, Bologna, pp 811–816

Zangerl R (1969) The turtle shell. In: Gans C, Bellairs Ad'A (eds) Biology of the Reptilia, vol 1. Academic Press, London, pp 311–339

Zug GR (1991) Age determination in turtles. SSAR Herpetol Circ 20:1–28

Zug GR, Parham JF (1996) Age and growth in leatherback turtles, *Dermochelys coriacea* (Testudines: Dermochelyidae): a skeletochronological analysis. Chelon Conserv Biol 2:244–249

Systematic Index

Subject Index

The World Biodiversity Database

Carl H. Ernst, Roger W. Barbour

Turtles of the World

CD-ROM. 1998.
Windows: ISBN 3-540-14547-8
Macintosh: ISBN 3-540-14548-6
DM 228,-

In cooperation with the Smithsonian Institution, University Press and ETI, Professor Ernst has updated and extended his excellent book on turtles with many unique colour, and black and white photographs. About 40 new species have been added and the taxonomy completely revised. The geographic information system, MapIt™, includes new distribution maps allowing the user to determine easily in which parts of the world different species live, and which live in your area. An illustrated, interactive identification key allows the user to identify the turtles quickly.

All text is hyperlinked: an illustrated glossary defines about 300 specific scientific terms. All principal literature references for this group are included in this program. This CD-ROM is an excellent introduction to this group of wonderful animals.

Please order from
Springer-Verlag Berlin
Fax: + 49 / 30 / 8 27 87- 301
e-mail: orders@springer.de
or through your bookseller

Errors and omissions excepted.
Suggested retail price plus local VAT.

Springer

Springer-Verlag, Postfach 14 02 01, D-14302 Berlin, Germany

Springer
and the
environment

At Springer we firmly believe that an
international science publisher has a
special obligation to the environment,
and our corporate policies consistently
reflect this conviction.

We also expect our business partners –
paper mills, printers, packaging
manufacturers, etc. – to commit
themselves to using materials and
production processes that do not harm
the environment. The paper in this
book is made from low- or no-chlorine
pulp and is acid free, in conformance
with international standards for paper
permanency.

Springer